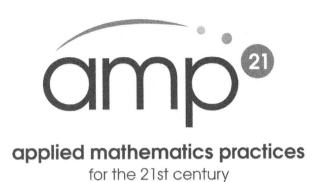

applied mathematics practices
for the 21st century

MATHEMATICAL MODELING WITH PROBABILITY

Using Authentic Problem Contexts

Kenneth Chelst, Ph.D.
Thomas Edwards, Ph.D.

applied mathematics practices
for the 21st century

Copyright © 2018 Kenneth Chelst and Thomas Edwards.
All rights reserved.

This book or any portion there of may not be reproduced or used in any manner whatsoever without the expressed written permission of the authors except for the use of brief quotations in a book review.

Applied Mathematics Practices for the 21st Century (AMP21) is an affiliation of professors and mathematics educators who share a common desire to bring relevance to K-12 mathematics by using authentic problem contexts in teaching and developing mathematics concepts and skills. The lead faculty are Kenneth Chelst (College of Engineering) and Thomas Edwards (College of Education) who are both professors at Wayne State University in Detroit, MI.

AMP21 is a Non-Profit developer and provider of curriculum that is aligned with the eight Standards for Mathematical Practice in the Common Core State Standards. Our team has published two textbooks for high school, 1) Mathematical Modeling with Algebra and 2) Mathematical Modeling with Probability. Our new curriculum development focuses on middle school topics central to proportional reasoning: percentages, rates, ratios and proportions. Everything is presented in authentic problem contexts that involve making decisions. AMP21 offers professional development workshops in conjunction with local universities to help teachers and schools develop the programs needed to enable students to succeed in the global economy. In addition, this curriculum can form a basis for project based learning in mathematics.

First Printing, 2018
ISBN -1986267733

Applied Mathematics Practices (AMP21)
In conjunction with Wayne State University
www.appliedmathpractices.com

Contact: Kenneth Chelst
4815 Fourth St. (Room 2017)
Detroit, MI 48201
kchelst@wayne.edu

Acknowledgements:

This work was developed initially with funding from the National Science Foundation (NSF Directorate for Education and Human Resources Project # DRL-0733137). The grant was a partnership between North Carolina State University, University of North Carolina – Charlotte, and Wayne State University. The following people contributed to the initial project, known as Project MINDSET:

Wayne State University

Dr. Thomas Edwards
Dr. Kenneth Chelst
Mr. Jay Johnson

North Carolina State University

Dr. Karen Keene
Dr. Karen Norwood
Dr. Robert Young
Dr. Molly Purser
Dr. Amy Craig-Reamer

University of North Carolina Charlotte

Dr. David Royster
Dr. David Pugalee
Ms. Sarah Johnson

Researchers

Dr. Saman Alaniazar
Dr. Hatice Ucar
Dr. Richelle Dietz
Dr. Krista Holstein
Mr. Will Hall
Ms. Zeyneb Yurtseven

Teachers

Derek Blackwelder, Jamie Blanchard James Bruckman, Kevin Busfield, Robert Butler, Molly Charles, Suzanne Christopher, Kristina Clayton, Krystle Corbin, Dean DiBasio (teacher writer), Robin Dixon, Caroline Geel, Michael Gumpp, Breanna Harrill, Jack Hunter, Amy Johnson, Jennifer Johnson, Jenise Jones (teacher writer), Anna Kamphaus, Chris Kennedy, George Lancaster (teacher writer), Craig Lazarski, Jeffrey Loewen, Charles Ludwick, Margaret Lumsden, Leon Martin, Cora McMillan, Kristen Meck (teacher writer), Karen Mullins, Erica Nelson, Bethany Peters, Joseph Price, Angela Principato (teacher writer), John Pritchett, Tyler Pulis, Paul Rallo, Tonya Raye, Kym Roberts, William Ruff, Amy Setzer, Rob Sharpe, Christy Simpson, Catherine Sitek, Cheryl Smith, Julia Smith, Wendy Carol Srinivasan, Susan Sutton, Greg Tucker, Laura Turner, Thad Wilhelm (teacher writer), Tracey Walden, Tracey Weigold, Lanette Wood

Cover Design: Susan Marie Dials

Significant contribution to the 2018 second edition revision: Thad Wilhelm

MATHEMATICAL MODELING WITH PROBABILITY
Using Authentic Problem Contexts
Kenneth Chelst and Thomas Edwards

Table of Contents

Introduction to Managing and Decision Making in an Uncertain World	vii
Chapter 1: Basic Probability and Randomness	3

- Super Bowl • simulate randomness • customer service
- late newspaper • absenteeism

Chapter 2: Conditional Probability	71

- mortality • committees • basketball free throws • home and road games
- faulty ignitions • graduation rate • celiac disease

Chapter 3: Decision Trees	125

- prom location • automation • energy plant • collision insurance

Chapter 4: Binomial and Geometric Distributions	167

- customer service • incomplete newspaper • absenteeism
- blogger • customer service • NASA shuttle

Chapter 5: Poisson Distribution	229

- maternity ward • CSI team • health care clinic

Chapter 6: Normal Distribution	265

- parachute fabric • battery warranty • toy demand

Chapter 1: Basic Probability and Randomness

We all face decisions in our jobs, in our communities, and in our personal lives that involve uncertainty. When making such decisions, there can be no guarantee that the outcome will be favorable. Probability decision models are designed to help an individual to assess the likelihood of various outcomes. The decision maker then uses that information to select a preferred choice that explicitly considers the relative likelihood of both positive and negative results.

In perceiving and responding to the world around us, it is critical to develop an understanding of the difference between patterns of random and non-random events. There is well documented fallacy in probabilistic thinking in which people see patterns when in fact all they have observed is random fluctuations. A decision maker faced with a series of negative outcomes must decide whether or not to act. Could this simply be a random fluctuation and no action needs to be taken? Or is there something causing these negative outcomes that he can and should address?

In this chapter we strive to develop a better understanding of random patterns through the use of simulated experiments. We begin with a physical simulation, flipping a coin. This can be used to model randomness with two equally likely events. Random number generators are introduced to represent randomness when two events are not equally likely. These random number generators are found in advanced calculators and Excel spreadsheets.

This chapter introduces the basic concepts and notation of probability. The multiplication rule is used to calculate the probability of occurrence of two independent random events. The principle of complementarity is used to indirectly compute the likelihood of an event. Missing from this chapter is any discussion of counting methods, combinations and permutations. Historically, these were developed primarily to apply probability to games of chance. In our experience we have never had to use counting rules to tackle non-gambling decisions involving uncertainty.

Chapter 2: Conditional Probability

Chapter 2 introduces the concept of conditional probability, $P(B|A)$. Given that event A has occurred, what is the probability that event B will occur? Unlike most probability textbooks, we dedicate an entire chapter to explore conditional probability. Research has shown that many people have poor intuition regarding conditional probability concepts.

This chapter illustrates how to determine conditional probability from data tables as well as from a problem context. It illustrates the application of the multiplication rule to calculate joint probabilities of two events that are not independent. The chapter proceeds to develop the formula for a probabilistic partition. In this chapter we introduce the concept of a random variable. This is a function that translates the outcome of a random experiment into a unique numeric value. The measure of central tendency of this random variable is a probabilistic weighted sum called the expected value.

Chapter 3: Decision Trees

Decision trees provide a structure for determining the alternative that optimizes the expected value. Decisions are represented by rectangles and uncertain events by ovals. Branches emanating from a rectangle correspond to distinct decision alternatives. Branches from a random event correspond to different possible outcomes. The decision tree also provides the probability distribution for each of the alternatives.

The first example simply introduces the concept of tradeoff between more or less conservative decision alternatives and possible outcomes. Two subsequent decision contexts maximize the expected profit. The last example minimizes the expected cost of collision insurance.

Chapter 4: Binomial Distribution and Geometric Distribution

In Chapter 1 we used basic probability and simulation to study random events with two mutually exclusive outcomes: answer or do not answer phone, at work or absent from work, on time or late. In chapter 4 we introduce the concept of a probability distribution function. When repeated random events follow the same assumptions, there may be a formula that summarizes the probabilistic pattern. This formula will have parameters whose values vary from context to context.

The binomial distribution can be applied to a situation of identical independent repetitions of the same random experiment. The four common elements to the random experiment are:
1. only two possible outcomes: success and failure
2. p, the probability of success, is the same for each repetition
3. each repetition is independent of every other repetition
4. n identical repetitions of the random event

The parameters of the binomial distribution are n and p. The formula for the binomial distribution calculates the probability of X successes out of n trials. The random variable X has a range from zero to n. Statistical calculators and Excel spreadsheets include formulas for the binomial distribution and the cumulative binomial distribution. The binomial distribution examples in this chapter are all extensions of those that appeared in Chapter 1.

In the binomial distribution we count the number of success out of n trials. The geometric distribution includes the same first three elements as the binomial. However, the random variable that is tracked is the number of repetitions until the first success. This random variable can take on the values of one to infinity. One interesting application of the geometric distribution involves the number of shuttle flights until the first shuttle disaster.

Chapter 5: Poisson Distribution

The Poisson distribution is used to characterize a probabilistic environment in which random events occur totally independent of one another. Emergency calls to 911 or calls to a telephone helpline are examples of situations that have been modeled with the Poisson distribution. This distribution has one parameter, λ, the average number of incidents per unit of time. The distribution is used to estimate the probability that there will be X events in a unit time. Alternatively, it can be used to model an extended time period, t. In that case, the average or expected value of the random variable is λt. The random variable X is discrete and has a range of zero to infinity. Managers face the difficult task of scheduling the necessary resources to handle the fluctuations in workload caused by the unpredictability of these random events.

Chapter 6: Normal Distribution

The Normal distribution is used to describe a continuous random variable with a mean of μ and a standard deviation of σ. The sum of identically distributed independent random variables is known to approach the normal distribution. For that reason, the randomness of the sample average of a random variable can be approximated with the normal distribution. In the first example, the normal distribution is used to model the random error associated with cutting panels of material for use in constructing a parachute. The normal distribution can help determine the proportion of panels that are within specifications. The second example uses the normal distribution to help establish an appropriate warranty replacement policy. In the final example, a manager must decide how many units of a new toy to stock when the demand is normally distributed. The decision maker balances the risk of having too few items and losing out on sales against the risk of overstocking and having to steeply discount toys that are left over after the buying season passes.

Introduction to Managing and Deciding in an Uncertain World

The world around us is filled with uncertainty, risk, and variability that complicate day-to-day decisions and the development of long-term plans. This applies to personal decisions as well as decisions made by companies and government agencies. An applicant to college cannot be sure which schools will accept him. In a rush to get a meal between classes, a student faces uncertainty about the time it will take to be served at the school cafeteria. The school newspaper editor is concerned about how many members of the writing team will meet their deadlines. While reviewing alternative car insurance plans, the student driver struggles to decide on the size of the collision insurance deductible. Obviously, no one plans to have an accident, but the risk of an accident is always present. Companies that provide insurance look at the same problem. They come up with pricing strategies for insurance that pool the risk of an individual with large groups of similar people.

> *Variability* is a characteristic of data that refers to the recognition that each data value is not the same. For example, there is variability in height of individuals or in their annual income. Two common statistics used to characterize a dataset's variability are variance and standard deviation.

Companies launching a new product or service must deal with uncertain demand. Police patrol supervisors must consider random fluctuations in the demand for service and patterns of crime. Plant managers and school officials must cope with workers who randomly do not show up for work due to illness.

Probability theory and sophisticated probability models enable us to identify individual occurrences that are more or less likely, as well as estimate long-term averages. This knowledge and understanding is essential for planning and making critical decisions in advance as well as adjustments on the spot. A plant or office manager must decide how many spare workers to have on duty to address the problem of absenteeism. The high school newspaper editor will also have to plan what to do if only eight of 10 writers have completed their assignments. Similarly, an automobile dealership needs to know how much inventory to carry. The police commander must schedule the number patrol officers on each shift and where to place them without knowing where and when the next crime will occur. Often, decision makers must balance the added costs of having reserves against the impact of running short.

This text is designed to provide decision-making guidance in the presence of uncertainty. One of the challenges in learning basic concepts of probability is that many of us do not have good intuition about randomness. We will address this problem while developing probabilistic decision-making skills. The text is therefore designed to develop your ability to recognize and understand patterns of random events. We will do so by having you simulate random experiments first with a coin flip. Then, you will simulate more complex situations with the random number function in your calculator. Lastly, you will use the random number generator in Excel to develop and analyze a large sample of randomly generated data.

In introducing basic concepts, we routinely use the concept of *relative frequency* as an estimate of probability. Thus, our introductory examples will use data rather than the counting methods that you may have seen in other probability courses.

> ***Randomness*** is a characteristic of a process, experiment, or environment in which outcomes cannot be predicted with certainty. A random experiment, such as rolling a die, can yield different unpredictable outcomes. The gender of newborn is a result of a random process involved in becoming pregnant. The temperature on any day reflects randomness of our environment.

In later chapters we introduce mathematical formulas that can be used to describe different patterns of randomness. These are probability distributions. These formulas will be applied to two different types of variables: discrete and continuous. A ***discrete variable*** is countable. For example, the number of crimes in a day and the number of people absent from work are discrete variables. On the other hand, a ***continuous variable*** cannot be counted and is measured instead. For example, the time to complete an exam and the height of a randomly selected individual are continuous variables.

CHAPTER 1:

Basic Probability and Randomness

Section 1.0 Basic Probability and Randomness

In order to fully understand probabilistic modeling, you will need a good sense of the nature of *random* behavior. All of us have basic intuition about probability and randomness. This intuition develops over time from our experiences. These include what we watch on television, observe at sporting events, and read on the Internet. The problem is that research has shown that our intuition about randomness and probability is often flawed. For example, a manager being told that each and every one of his 10 suppliers is 95% reliable often believes the system he has in place is in good shape. In this chapter, you will learn why the manager needs to be concerned.

Because of the many misconceptions, the development of probability analysis skills requires demonstrating the flaws in our intuition. Only after this is accomplished, can you move on to learn formal approaches to probability. The first example we explore involves understanding what is often our preconceived notion of a random pattern. Specifically, we will ask you to imagine and write down a typical sequence of heads and tails when flipping a coin. You and your classmates will then be asked to actually flip a coin. By comparing the lists, we hope to dislodge a misconception you may have about random events.

One of the primary goals of this section and all future sections is to begin understanding how much **variability** there may be in a simple random process. This will be accomplished by comparing and contrasting the results of your coin flipping experiment with those of your neighbors and those of the class as a whole. For example, we will explore how much difference there is in the percentage of heads and tails in your list and those of each and every one of your classmates. We will then look at the average for the entire class.

One challenge managers and we personally face in a variety of situations involves understanding and interpreting fluctuations around a long-term average. Is a rise in absenteeism or lateness simply a reflection of a random fluctuation or an indication of a developing problem that needs to be addressed? Biostatisticians regularly are on the lookout for occurrences of disease that are out of the "normal" range. A report comes in that a small city has twice the national average rate of a specific type of cancer. There are tens of thousands of communities. Is it likely or unlikely purely by random fluctuation to see one or more cities with a rate twice the national average? If it is likely, no action should be taken, no detailed study initiated. If however, probability theory suggests this high rate for a small city is extremely rare, public health officials may commission an expensive study to gather more data to try to identify the contributing factors.

You will explore the pattern of which conference won the Super Bowl in an attempt to draw a conclusion about whether the data suggest that one conference is superior to the other. You will also look at data for a customer service telephone line to determine whether there is sufficient evidence that the service is not meeting the timeliness standard established by management. You will assist the faculty advisor to the school newspaper to better understand why they are having problems publishing the paper on time. Finally, you will assist a plant manager to decide how many spare workers to have on duty. These are needed to maintain productivity while coping with random fluctuations in the number of workers who are absent from work each day.

A key question you will always be addressing is, "What is the source of variability?" The fluctuation around the mean may be merely random variation or it could be the result of some external influence. Discerning whether the pattern of variation is random or due to some factor we might be able to control is a critical component of probabilistic and statistical analysis. We need these tools because, in trying to make this distinction, our own intuition often leads us to unfounded conclusions.

Section 1.1 Super Bowl – Conference Dominance

Every year since 1967, the American Football Conference (AFC) and National Football Conference (NFC) champions of the National Football League compete in a championship game known as the Super Bowl. Table 1.1.1 identifies by conference the winner of each of the 48 Super Bowls from I through XLVIII. The question is:

Super Bowl					
Year	Winning Conference	Year	Winning Conference	Year	Winning Conference
1967	NFC	1983	NFC	1999	AFC
1968	NFC	1984	AFC	2000	NFC
1969	AFC	1985	NFC	2001	AFC
1970	AFC	1986	NFC	2002	AFC
1971	AFC	1987	NFC	2003	NFC
1972	NFC	1988	NFC	2004	AFC
1973	AFC	1989	NFC	2005	AFC
1974	AFC	1990	NFC	2006	AFC
1975	AFC	1991	NFC	2007	AFC
1976	AFC	1992	NFC	2008	NFC
1977	AFC	1993	NFC	2009	AFC
1978	NFC	1994	NFC	2010	NFC
1979	AFC	1995	NFC	2011	NFC
1980	AFC	1996	NFC	2012	NFC
1981	AFC	1997	NFC	2013	AFC
1982	NFC	1998	AFC	2014	NFC

Table 1.1.1: Super Bowl winners by conference

What is it about the sequence of conference wins that has led most sports writers to believe that one conference or the other was superior at different extended periods in time?

1. What do you notice about these results?

2. Does one conference appear stronger than the other? Why or why not?

3. If neither conference is superior to the other, what number would it make sense to use as the probability that the winner will come from a given conference?

4. What was the overall percentage of wins by each conference? Each column represents a 16-year period. What was the winning percentage in each 16-year period?

In order to explore your conception of randomness, we will model the Super Bowl results since 1967. Assume that the two conferences are equally strong. If so, each conference winner has the same 50% chance of winning the Super Bowl in any year. We will use a coin flip to represent the 50% chance. First you will imagine and write down a typical sequence of heads and tails. Only afterwards, will you actually flip a coin.

5. On a sheet of paper, list the numbers from 1 to 48 to represent each of the Super Bowls from 1967 to 2014. Table 1.1.2 shows how to set up the list before entering the Hs and Ts. You will use an H to represent an AFC win and a T for an NFC win. Now imagine a random sequence of wins assuming the two conferences are equally strong. Record next to each number, an H or a T.

1		17		33	
2		18		34	
3		19		35	
4		20		36	
5		21		37	
6		22		38	
7		23		39	
8		24		40	
9		25		41	
10		26		42	
11		27		43	
12		28		44	
13		29		45	
14		30		46	
15		31		47	
16		32		48	
Number of Heads		Number of Heads		Number of Heads	
Percent of Heads		Percent of Heads		Percent of Heads	

Table 1.1.2: A format for recording simulated coin flips

After recording all 48 Hs and/or Ts, fill in the bottom rows of the table: the number of heads and percent of heads in each column. Also calculate the percent of Hs for all 48 imagined events.

6. Compare the percent of heads in your three columns and the overall percent. How much difference is there among the column percentages?

7. Compare the percent of heads in your list with those of two or three other students sitting close by. Are your results similar, or is there a lot of difference among them?

Chapter 1 Basic Probability and Randomness

Next, make another list just like the one in Table 1.1.2 to record the results of your actual coin flips. After recording all 48 Hs and/or Ts, fill in the bottom rows of the table: the number of heads and percent of heads in each column. Also calculate the percent of Hs for all 48 flips.

8. Compare the percent of heads in your three columns and the overall percent. How much difference is there among the column percentages?

9. Compare the percent of heads in your list with those of two or three other students sitting close by. Are your results similar, or is there a lot of difference among them?

In the following questions, you will compare your imagined sequence with the sequence of the actual coin flips. Be sure to label your lists, so that you'll know which is which.

10. Was there more variability in the column percentages among the imagined lists or the coin flip lists?

11. What were the maximum and minimum percentages for you and your neighbors for each column for the imagined lists and coin flip lists? Are you surprised at how far these numbers are from 50% in the actual coin flip data?

12. What were the maximum and minimum percent of Hs and Ts for you and your neighbors for all 48 imagined events and coin flips? What differences do you notice about the ranges?

13. For the entire class, what was the minimum percent of Hs observed when flipping the coins 48 times? What was the maximum? Are you surprised at how far these numbers are from 50%?

14. Did anyone in the class have a percentage that low or that high in the imagined list?

15. For the entire class, what was the overall percent of Hs and Ts for the coin flip random experiment?

We will now return to the actual data for the Super Bowl in Table 1.2.1. Find the total the number of times in each of the three columns that the NFC won and calculate the percent.

16. Look at the percentage of NFC wins in the first column of the actual Super Bowl data. In your coin flip table, was there any percentage of Hs or Ts that low? Did anyone in the entire class have a percentage of Hs or Ts that low in a column of coin flips?

17. Look at the percentage of NFC wins in the second column of the actual Super Bowl data. In your coin flip table, was there any percentage of Hs or Ts that high? Did anyone in the entire class have a percentage of Hs or Ts that high in a column of coin flips?

18. Complete the Table 1.1.3 for the class as a whole. In this table we want the frequency of either a large number of Hs or Ts. In the summary of the 16 year periods, we would be surprised if either of the divisions won an unusually large proportion of times. Thus, in assessing the likelihood of one conference winning, for example, 11 or more times in 16 years, we need to consider both Hs and Ts.

	Large number of Hs		Large number of Ts		Hs and Ts Combined	
Number of Hs or Ts	Frequency	Relative Frequency	Frequency	Relative Frequency	Frequency	Relative Frequency
11						
12						
13						
14						
15						
16						

Table 1.1.3: Class frequency of high percentages

19. Based on Table 1.1.3, what is the likelihood that one conference would win 11 or more Super Bowls in a 16 year period? 12 or more? 13 or more? 14 or more?

20. Based on your comparisons between your actual coin flips and the Super Bowl wins, do the percentages strongly support the news writers' claims of dominance in any of the 16 year periods?

Up to this point we have explored different aspects of the percent of Hs and Ts. We have compared the imagined sequence with that generated by actual coin flips. We have also compared minimum and maximum percentages for columns and for the total. By including your classmates, you were able to consider larger and larger sets of data. For the class as a whole, you should have seen a wider range of percentages for the column percentages than you found in your data and your neighbor's. In addition, there should have been less variability in the total percent of conference wins over 48 years. The total percent for the entire class would likely be close to 50%.

Now we are going to look at the sequences of consecutive Hs or Ts. These sequences are called strings. There have been numerous classroom experiments comparing made up lists and randomly generated lists. In general, the lists of imagined sequences of Hs and Ts tend to have shorter strings than randomly generated sequences. In addition, imagined lists rarely have any long strings of four more. We are going to see if your experience and those of your classmates replicate the research.

Record the lengths of every string of consecutive heads or tails in your imagined list. Then do the same for your coin flip list. For example, in Table 1.1.4, the first column records a sequence of Hs and Ts. The second column records the length of each string. In Table 1.1.4 the longest string is 3.

T		
T		2
H		1
T		1
H		
H		2
T		1
H		
H		3
H		
T		
T		3
T		
H		
H		2
T		1

Table 1.1.4: Length of strings for imagined list

21. Compare the lengths of the strings in your two lists. Which list has the longest string? How long is that string? Compare your answers with those of two or three neighbors. In your group, what is the length of the longest string? Which list did it come from?

22. Compare your answers with the entire class. What is the length of the longest string? Which list did it come from?

23. In the Super Bowl results, what was the longest string of wins for one conference? Did any of the coin flip lists have a string this long?

24. Based on your comparisons do the data support the sports writers' claims of dominance?

25. Have you discovered any misconceptions in your own thinking about randomness?

1.1.1 Percentage, Long Strings, and Probabilistic Thinking

Every alternative sequence of 48 heads and tails has exactly the same probability of occurrence, one in 281.5 trillion. Thus the likelihood of seeing any particular sequence of alternating heads and tails is the same as the probability of seeing a sequence of all heads or all tails. However, the likelihood of producing exactly 24 heads out of the total of 48 is orders of magnitude higher than the chance of 48 heads in a row. That is due to the fact that there are many sequences of heads and tails that result in exactly 24 heads, but there is only one way to obtain 48 heads in a row. Thus, 48 heads in a row might lead you to believe that the coin is two-headed.

This same principle points to an interesting phenomenon. When random flips of a coin are carried out, the sequence often includes multiple examples of long runs of four or more heads or four or more tails. However, individuals recording their own made-up sequence have a tendency to believe that randomness means frequently alternating back and forth between heads and tails. Consequently, they tend to create lists that have few long strings of heads or tails. As a result, an experienced probabilistic thinker can often identify which of the two lists was generated by flipping a coin and which was generated by a human mind. The human mind tends to look at the list as it develops and work to balance the distribution of heads and tails.

> One of the primary goals of probabilistic analysis is to attempt to assess whether a pattern is simply random or if some factor is contributing to the pattern. It is never possible to resolve this issue with certainty, but an understanding of probability theory can help decide which driving force is more likely.

Section 1.2 Simulate Two-outcome Random Event

On the pages that follow, you will be introduced to a wide range of real-world problems that operations researchers routinely encounter. They approach these real-world problems using an array of techniques that apply principles of probability and statistics. You may already be familiar with many of these concepts, but some of them might be new. We believe that a major difference between our use of probability and statistics in this textbook and what you may have seen in the past is that we will use probability and statistics in the context of making decisions.

Along the way, you will also learn about the role of computers in exploring real-world problems by actually using computers. A computer allows us to simulate a situation rather than attempt to collect large amounts of data. It can also enable us to realistically model changes in the design of the system and determine the impact on system performance. This is much preferred to trying out each change in the real world and observing the impact. When we use mathematics in this way, it is more as a modeling tool and less as a computational tool. The primary role of an operations research (OR) professional is to formulate the model and interpret the results that the model returns.

Studying mathematics in this way may require you to develop a new mindset about what mathematics is, how it can be used, and the best ways to learn it. Whatever mathematics is, it is certainly not a spectator sport. That is why a large portion of this book looks very similar to the Super Bowl simulation problem that you just read. Many questions were asked, but few were answered in the text. Where will the answers come from? You will provide those answers with the help of your classmates and teacher. In doing so, you will also discover that applying mathematics to solve real-world problems can be a relevant, creative, and exciting team sport. We hope that all of this will help you begin to form that new mindset.

1.2.1 Advance Planning

Imagine you and your extended family of 50 people have planned a major outdoor dinner barbecue. Do you have a plan for rain? What if a week in advance you are told there is a 5% chance of rain? What if it is 50% or 80%? What options do you have if you start planning a week in advance that you may not have the morning of the event? What communication challenges do you face? How can you plan for them to ensure as much as possible that everyone knows what to do? The situation a manager faces is similar. When he is trying to schedule work to be completed, he is not sure how many workers will be absent on any given day. He also cannot be sure how many of them will complete their assigned tasks on time. The more he understands about random phenomena, the better able he is to develop a plan to deal with these uncertainties.

1.2.2 A Simple Uncertainty Model – Two Possible Random Outcomes

The simplest kind of uncertainty to model is one that involves only two possible equally likely outcomes, such as a coin flip. There are many other uncertain situations with only two possible random outcomes: a worker or student is present or absent, an assignment is handed in on time or not, or a team wins or loses a game. However, in these cases the two possible outcomes are not necessarily equally likely. The critical number, often called a **parameter**, is the probability of the outcome occurring. For example, it would be 0.1 if there is a 10% chance a worker will be absent on any given day. It might be 0.95, if a student is extremely reliable in turning in his or her assignments on time. In both cases, the pairs of outcomes are both *mutually exclusive* and *complementary*. They are mutually exclusive, because both outcomes cannot happen at the same time. A flip of a coin is either heads or tails but not both. They are complementary, because if one does not happen, the other must. If the coin does not show a head, then it must show a tail. If a worker shows up for work, then he is not absent and vice versa. Complementary events are two events that are also *collectively exhaustive*. The two complementary events exhaust all of the possible outcomes. As a result, the sum of the two probabilities must equal one. The equation below uses mathematical notation to express this complementary relationship.

$$P(E) + P(E^c) = 1$$

Mathematical notation is used to concisely express ideas about quantities. However, it is difficult to interpret if you do now know what the symbols mean. In probability theory, a capital P followed by an event enclosed in parentheses represents the probability of that event occurring. So $P(E)$ represents the probability that event E will occur. The symbol E^c represents the complement of event E.

This leads to a useful formula for calculating the probability of the complement of an event.

$$P(E^c) = 1 - P(E)$$

This simple situation forms the basis for all of the examples in this chapter. We expand on this two-alternative context to study what happens when the same situation is repeated three, five, 10, 20, or more times. Here we are interested in the total number of times one outcome or the other occurs. In the worker situation, we are interested in how many workers are absent each day because management is concerned about the potential impact on productivity.

The starting point in modeling these uncertain situations is to determine the probability, p, of a specific occurrence. We arbitrarily call this a success. For a coin that is fair, we assume there it is a 0.5 probability of a head and a 0.5 probability of a tail. For the worker case, the manager would look at historical data and determine the proportion of days missed by the workers in his organization. The challenge is to use this value to understand what happens when the same situation is repeated multiple times. For example, how many heads will occur if we flip a coin 12 times? It is extremely unlikely if we repeat the 12 coin flips two, three, or four times that we will see exactly the same sequence of heads and tails. However, we will explore what types of predictions we can make, and with what degree of likelihood.

Anytime we gather data about an uncertain situation, we have information about how things were. What we do not know is how likely this situation is to be repeated. What happens if we change something, what will the random patterns look like then? For example, our small office of 10 workers may have extensive data on the random number of workers who are absent from work each day during flu season. How would we predict from this information what the situation is likely to be if we expanded to 14 workers or cut back to seven? A mathematical probability model can assist us in this analysis.

1.2.3 Simulations

The simplest and most flexible method of modeling probability is to *simulate* the situation. For example, assume there is a 50% probability of reaching a customer service representative right after opening time We could let "heads" represent reaching a customer service representative and "tails" represent not reaching a representative. We can simulate a work-week by flipping the coin five times and recording the results. Then we can count the number of times in the week that customer service was open on time. In this way, by flipping a coin, we can get a sense of what the results of actually calling might have been.

A 50-50 probability can be simulated by flipping a coin, but what about a random event in which the chance of success is 0.7 and the chance of failure is 0.3? In this chapter we will use the random number generator found on your calculator and also in EXCEL to simulate these types of situations. To simulate a 50-50 probability, we can generate any two random integers to represent the two equally likely outcomes. When the two events are not equally likely, we need to be more creative. To simulate a 0.7 chance of success, we will generate random integers between 1 and 10. If the number is 7 or less, we record a success and if the number is 8 or more, we record a failure. To simulate a 0.95 chance, we will generate random integers between 1 and 100. Any value less than or equal to 95 represents a success. If the probability is recorded to three decimal places, we will use the integers from 1 to 1,000. With this simple modeling tool we will be able explore, study, and come to understand two-outcome random phenomena with different values of p, the probability of success.

Section 1.3 Customer Service at Koala Foods

Like most companies, Koala Foods, a wholesale distributor of foods to supermarkets, is concerned about its customer service. Koala Foods assures the supermarkets it serves that their needs can be addressed on weekdays beginning at 8:00 a.m. by calling a toll-free customer service hotline. The president of Koala Foods wants to increase customer satisfaction, so he hires a company that will monitor the customer service department. Specifically, the monitoring company will initially check daily that the customer service department is actually operational by 8:00 a.m. Later the monitoring company will track individual calls to assess the quality of service.

1.3.1 Forming a Strategy

AGB Company, a customer service consulting firm, has been contracted by Koala Foods to monitor its customer service department. There are various ways of measuring quality in this situation. Mr. Smith, the President of Koala Foods, proposes checking to see if the customer service department is operational by 8:02 a.m., at 8:05 a.m., and lastly at 8:10 a.m. He would be satisfied if the customer service department was ready and answering calls by 8:02 at least 50% of the time. He sets a higher standard of 70% for 8:05 a.m. He feels that by 8:10 a.m. the service should almost always be open and sets a standard of 90%. This is equivalent to the service being ready to answer calls at 8:10 a.m. 9 days out of 10.

On the first three days of the week, AGB Company calls the Koala Foods customer service department at 8:02 a.m. Each day they encounter an "hours of service" recording rather than an actual customer service representative.

1. Does Koala Foods have a problem?

2. What are some possible reasons a call would not be answered at 8:02 a.m.?

3. Is this enough information to decide that there is a problem in meeting the 50% standard?

1.3.2 Simulating the Situation

We are interested in better understanding the significance of failing to reach customer service three days in a row. We need to decide whether this is sufficient evidence of a problem. If so, Koala Foods management will have to introduce greater supervision to ensure the standard is met. The alternative is to continue calling each morning to collect more data to see if there truly is a problem.

In order to gain the necessary understanding of this sequence of three consecutive failures, AGB Company decides to simulate three consecutive calls a large number of times. They want to know how likely it is to see three failures in a row if the 50% standard was being met overall. You and your classmates will assume the role of AGB and simulate this 50-50 random experiment by flipping a coin. Let the "heads" side of the coin represent a call that gets through to a customer service representative. The "tails" side of the coin represents a call that receives a recorded message.

Let   H = answered call
      T = unanswered call (recorded message)

4. Do you think that your first three consecutive flips will all be tails? If you repeat a set of three flips a total of 10 times, do you think you will ever see three tails in a row?

5. Do you think that anybody in the class will ever get all three tails?

6. Flip the coin three times and record the results. Repeat this process nine more times. How many times did you observe a series of three tails?

7. Did any of your classmates flip a series of three tails? Count the total number of times that you and your classmates observed a set of three tails in a row. What is the proportion of sets of three flips that were three tails in a row?

8. Think back to Koala Foods Customer Service Department and meeting the 50% standard. Does it seem unusual that AGB failed to reach customer service at 8:02 a.m. three days in a row if the 50% standard was being met?

Another method of simulation uses the random number generator in your graphing calculator. It is capable of simulating three repetitions of this random event with one single command. It is also very easy to repeat this process as many times as you want. Within the MATH menu, the PRB submenu contains the random integer command, randInt(. If you enter randInt(0,1,3), your calculator will return a set of three random integers that are either "0" or "1." Each time you push ENTER, your calculator will generate and display a new set of three values. Let "0" represent a call that receives a recorded message and "1" represent a call that gets through to a customer service representative.

9. Enter randInt(0,1,3) into your calculator ten times and count the number of times the string {0 0 0} occurs.

10. Compare your answer to #9 with those of your classmates.

11. Would you reconsider your answer to #4 and #5?

12. Should you reconsider your response to #1?

After seeing the results of the simulation, Mr. Smith decides the data presented by AGB Company was inconclusive. He now knows that he would need data for more than three days to conclude with confidence that the standard of 50% is not being met.

1.3.3 Probability and Relative Frequency

The relative frequency of an event in a large number of simulations can be used to estimate the underlying probability of the event happening. However, by making some basic assumptions, it is often possible to calculate the probability directly. This is true for the coin flipping example. Each time a fair coin is flipped, the chance of observing a head or tail is 0.5. It is independent of the previous outcome. The coin has no memory, so it does not "remember" the result of the previous flip. The probability that a set of *independent* events will occur can be found by multiplying the probabilities of each of the events.

> The *relative frequency* of an event is equal to the ratio of the number of times the event occurred to the total number of observations.

In this example, we are interested in the likelihood of observing three tails in a row. Using the multiplication rule, this probability is

$$(0.5)(0.5)(0.5) = 0.125$$

or

$$\left(\frac{1}{2}\right)\left(\frac{1}{2}\right)\left(\frac{1}{2}\right) = \frac{1}{8}.$$

The likelihood is one chance in eight. As a result, most of the students who carried out 10 repetitions of three coin flips would have observed at least one set of three tails. Earlier, the students were asked to accumulate the class's data to determine the overall relative frequency of three tails. The relative frequency should be close to 0.125, but not exactly that value because of randomness.

Thus, it is possible that Koala Foods is maintaining a 50% standard, but that these particular three days represent an unusual but not rare event. The standard that is often used to characterize an event as rare is less than one-in-twenty or 0.05. In some instances an even stricter standard of one-in-a-hundred, 0.01, is used to characterize something as a rare event.

> A *random experiment* is any activity under consideration in which the outcome is unpredictable, such as rolling a die or checking who is absent from class. Each possible observation in an experiment is called an *outcome*. The set of possible outcomes of an experiment is called the *sample space*. Any subset of a sample space is called an *event*.
>
> A *simple event* is an event that cannot be decomposed. For example, 1, 2, 3, 4, 5, and 6 are all simple events associated with rolling a die once. The sample space in this experiment is {1,2,3,4,5,6}. Rolling an even number is a *compound event* that is made up of three simple events {2,4,6}.

In this example we used the flip of a coin to represent two equally likely outcomes. The sample space for the coin experiment consists of all possible outcomes of flipping a coin three times. The set of outcomes is listed below.

{H,H,H}
{H,H,T}
{H,T,H}
{T,H,H}
{H,T,T}
{T,H,T}
{T,T,H}
{T,T,T}

All of these eight outcomes are equally likely. As a result, the likelihood of any particular outcome, such as {T,T,T}, is 1/8.

In general, when all outcomes are equally likely, we can use the following formula to calculate the probability of an event E occurring.

$$P(E) = \frac{\text{Number of ways event } E \text{ can occur}}{\text{Number of possible outcomes}}$$

We can use this to determine the likelihood that on only one day out of three the phone is answered at 8:02 a.m.

Let E_1 = the event that only one call is answered and two calls are unanswered at 8:02 a.m.

This is a compound event corresponds to observing two tails and one head on three flips of a coin. The list of all possible ways of E_1 occurring consists of the following outcomes.

{H,T,T}   {T,H,T}   {T,T,H}

Thus, out of the eight total possible outcomes, there are three distinct, mutually exclusive, and equally likely outcomes that correspond to event E_1.

$$P(E_1) = \frac{3}{8} = 0.375$$

1.3.4 Dealing with a Higher Standard

Recall that the standard for reaching customer service at 8:05 a.m. was 70%.

13. Does the 70% standard for calls placed at 8:05 a.m. seem like a reasonable expectation if you are a
    a. manager at Koala Foods?
    b. customer service representative at Koala Foods?
    c. customer of Koala Foods who needs service?

AGB also called at 8:05 a.m. and three days in a row, no one answered. Mr. Smith wonders if there is any reason why these three days of data might be conclusive about whether the customer service department is meeting this higher standard. He decides to ask AGB to conduct another simulation. Once again, you and your classmates will play the role of AGB.

14. Will a coin be adequate to simulate this scenario? Why or why not?

To use the calculator, enter randInt(1,10,3). This will generate three random integers between 1 and 10, inclusive. Assign the numbers "1" through "7" to represent calls that are answered by customer service. The numbers "8" through "10" represent calls that receive machine recorded message.

15. Why does this method accurately model the 70% standard?

16. You and your classmates will perform this simulation 10 times each. Do you think there will be more, the same, or fewer simulated instances of receiving the recorded message three times in a row as compared to the first simulation?

Execute this simulation on your calculator 10 times. For each set of three, count and record the number of times a number between "8" and "10" occurs.

17. Did you have any lists of three numbers in which all three numbers were 8, 9, or 10?

18. Did any of your classmates observe a set of three numbers that were all 8, 9, or 10? Count the total number of times this occurred in the class's pooled data. In what proportion of total simulations did that occur?

19. Think back to Koala Foods Customer Service Department and meeting the 70% standard. Does it seem unusual that AGB failed to reach customer service at 8:05 a.m. three days in a row if the 70% standard was being met?

With the higher standard, the probability of answering a call is 0.7. Not answering the call is the complementary event, E^c. The probability of a complementary event is equal to one minus the probability of the event.

$$P(E^c) = 1 - P(E)$$

$$P(\text{Not answering call}) = 1 - P(\text{Answering call}) = 1 - 0.7 = 0.3$$

We can again use the multiplication rule to determine the likelihood of observing three days of non-answered calls at 8:05 if Koala Foods were achieving the 70% standard.

$$P(\text{No answer three days in a row}) = P(\text{No answer day 1}) \cdot P(\text{No answer day 2}) \cdot P(\text{No answer day 3})$$
$$= (0.3)(0.3)(0.3)$$
$$= 0.027$$

This is approximately one chance in 37. As a result, most of the students who carried out the 10 simulations would not have observed even one instance of a set of three numbers that were all 8, 9, or 10. Three days of unanswered calls is characterized as a rare event based on the 5% criterion mentioned earlier. Thus, it is reasonable to conclude that the customer service department of Koala Foods is not meeting the 70% standard for 8:05. Management will need to take action to correct the problem.

In calculating probabilities related to this event, we cannot apply the strategy of counting the number of possible outcomes that correspond to the event. All of the possible events ae listed below. No call answered three days in a row is just one of eight possible outcomes. However, its probability is not 1/8 or 0.125

Let A = answered call
       $A^c = N$ = not answered call

   {A,A,A}
   {A,A,N}
   {A,N,A}
   {N,A,A}
   {A,N,N}
   {N,A,N}
   {N,N,A}
   {N,N,N}

These are all of the eight possible outcomes. However, they are not equally likely. Consider the likelihood that the 8:05 call was answered three days in a row. We can calculate the probability with the multiplication rule for independent events. If we assume the 70% standard, the probability is shown below.

$$P(A, A, A) = P(A) \cdot P(A) \cdot P(A) = 0.7^3 = 0.343$$

This is very different from the probability of no calls being answered.

$$P(N, N, N) = P(N) \cdot P(N) \cdot P(N) = 0.3^3 = 0.027$$

Chapter 1 — Basic Probability and Randomness

1.3.5 Dealing with a 90% Standard

Recall that the standard for reaching customer service at 8:10 a.m. was even higher. It was set at 90%. AGB also called on Monday, Tuesday, and Wednesday at 8:10 a.m. This time the results were different. One out of the three times, they reached customer service. The other two times, they received a recorded message. Once again Mr. Smith wonders if these results indicate conclusively that the standard of 90% is not being met. He decides to ask AGB to conduct another simulation. Once again, you and your classmates will play the role of AGB.

20. Determine the beginning and the end of an appropriate random integer interval that could be used to simulate this situation. Assume again a three-day monitoring period.

    randInt(____,____,____)

21. What range of integers represents a call that is answered by a customer service representative?

22. What range of integers represents a call that receives a recording?

As you run your simulation and observe three numbers, you will need to record how many times a number representing an answered call appears. Run your simulation 10 times.

23. For your 10 simulations, how many times were 0, 1, 2, and 3 calls answered by customer service?

24. In your class, how many times were 0, 1, 2, and 3 calls answered by customer service?

Mr. Smith wonders whether only one answered call out of three calls is sufficient evidence to be concerned. If so zero out of three answered calls is even stronger evidence.

25. Did any member of the class observe either zero or one call being answered? How many times did this happen in the whole class? What proportion is this?

26. With regard to the 90% standard being met, what do you think Mr. Smith should conclude from AGB's three phone calls in which only one was answered at 8:10 a.m.?

In order apply probability to this context, we will introduce set notation and Venn diagrams. Let:

M = call answered on Monday at 8:10 a.m.
T = call answered on Tuesday at 8:10 a.m.
W = call answered on Wednesday at 8:10 a.m.

M^c = call NOT answered on Monday at 8:10 a.m.
T^c = call NOT answered on Tuesday at 8:10 a.m.
W^c = call NOT answered on Wednesday at 8:10 a.m.

To develop the analysis, we begin by first considering just two days, Monday and Tuesday. Each circle corresponds to the event that an 8:10 a.m. call was answered on that day. The intersection does not mean that the days overlap. It refers to the event that the phone call was answered at 8:10 a.m. on both Monday and Tuesday.

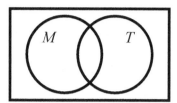

Figure 1.3.1: Venn diagram for calls 8:10 a.m. on Monday and Tuesday

The multiplication rule can be used to determine the probability of both events happening.

$$P(M \text{ and } T) = P(M \cap T) = P(M) \cdot P(T) = 0.9 \cdot 0.9 = 0.81 \quad \text{or} \quad \text{about 4 chances in 5}$$

Thus if the 8:10 a.m. standard were being met, there is more than 80% probability the phone would be answered on both days. The probability that the phone would be answered on at least one of the days corresponds to the **union** of these two events. The union of two sets, A and B, contains all the elements included either A or B. The formula for the number of elements in the union of two sets, A and B, is

$$|A \cup B| = |A| + |B| - |A \cap B|.$$

In this formula $|A|$ represents the number of elements that belong to set A. The logic behind this formula is that adding the elements of the two sets together involves counting all elements in the intersection twice. The intersection is part of both sets A and B. This formula directly translates into a corresponding formula for the probability that the outcome will fall within the union of two sets.

$$P(A \cup B) = P(A) + P(B) - P(A \cap B)$$

In this context, the formula for the probability of union of M and T is

$$\begin{aligned} P(M \cup T) &= P(M) + P(T) - P(M \cap T) \\ &= 0.9 + 0.9 - (0.9 \cdot 0.9) \\ &= 0.99 \end{aligned}$$

Thus, we calculate the sum of the probabilities of falling within each individual set and subtract the probability of the intersection. The likelihood, that the call will be answered on at least one day (only Monday or only Tuesday or both Monday and Tuesday) is extremely high, almost 100%.

The probability of a call not being unanswered on a specific day is (1 – 0.9) or 0.1. Thus, if the 90% standard were generally being met, it is extremely unlikely that they did not reach an operator at 8:10 a.m. on both Monday and Tuesday.

$$P(M^c \cap T^c) = P(M^c) \cdot P(T^c)$$
$$= (0.1)(0.1) \quad \text{or} \quad \text{1 chance in 100}$$
$$= 0.01$$

This value could have also been determined by using the concept of complementary events. The complement to the event that the call was not answered on both Monday and Tuesday is the event that the call was answered on at least one of those days. That probability of that event was determined to be 0.99. Thus

$$P(M^c \cap T^c) = 1 - P(M \cup T)$$
$$= 1 - 0.99$$
$$= 0.01$$

When tackling complex probability questions involving compound events, there is often more than one way to determine the probability.

Now let's return to the three-day situation. First we will review the two extremes: calls are answered every day and calls are not answered each and every day.

$$P(M \cap T \cap W) = P(M) \cdot P(T) \cdot P(W)$$
$$= 0.9^3$$
$$= 0.729$$

Thus, if the standard were generally being met, the likelihood that calls were answered on three consecutive days is 0.729. However, there is still a significant probability, (1 – 0.729 = 0.271) that this will not happen, that at least one call will be missed. In contrast, it is extremely unlikely for no calls to be answered on three consecutive days if the 90% standard was generally being met.

$$P(M^c \cap T^c \cap W^c) = P(M^c) \cdot P(T^c) \cdot P(W^c)$$
$$= (1 - 0.9)^3 \quad \text{or} \quad \text{1 chance in 1,000}$$
$$= 0.1^3$$
$$= 0.001$$

The original question asked was, what can be concluded if a call was answered on only one day out of three. For now, the only way to calculate the probability involves detailed calculation. The event of interest can occur in any one of three mutually exclusive ways. The call was answered on Monday and not on Tuesday or Wednesday, $M \cap T^c \cap W^c$. Alternatively, the call was answered on Tuesday and not on Monday or Wednesday, $M^c \cap T \cap W^c$. Finally, the call was answered on Wednesday but not on Monday or Tuesday, $M^c \cap T^c \cap W$. Because these patterns are mutually exclusive, the total probability equals the sum of the individual probabilities. The probability of each of these distinct outcomes is the same as calculated below:

$$P(M \cap T^c \cap W^c) = P(M) \cdot P(T^c) \cdot P(W^c) = 0.9(1-0.9)(1-0.9) = 0.009$$

$$P(M^c \cap T \cap W^c) = P(M^c) \cdot P(T) \cdot P(W^c) = (1-0.9)0.9(1-0.9) = 0.009$$

$$P(M^c \cap T^c \cap W) = P(M^c) \cdot P(T^c) \cdot P(W) = (1-0.9)(1-0.9)0.9 = 0.009$$

Thus, the probability of calls being answered on only one day is three times 0.009 or 0.027. This probability is less than 3% and is considered highly unlikely. If this were the pattern observed, the conclusion would be that Koala Foods is generally not meeting the 90% standard for 8:10 a.m. The variability does not appear to simply be the result of random fluctuation. These observations suggest that the variability is due to a systemic problem that the management of Koala Foods needs to address.

Section 1.4 Getting *The Lancer* to Press

Each month Al Mitchell, the faculty advisor for a school newspaper, oversees the production of the newspaper, *The Lancer*. He has 10 student writers, five of whom are editors and five are staff writers. In order to get to press on time, it is necessary that Al's students finish their articles by the required deadline. That is, he needs each and every student to meet his or her deadline.

1.4.1 The Problem

Over the years, Al has found that that student writers meet their deadlines only 95% of the time. On the surface, this does not seem bad, but *The Lancer* frequently goes to press late. He wants to change that. Al recalls something from his high school probability class called the **multiplication principle**: The probability that a set of **independent** events will occur can be found by multiplying the probabilities of each event occurring. Formally stated, the multiplication principle says that:

> If X and Y are *independent* events, then the probability of both X and Y occurring is equal to the product of the probabilities of X occurring and Y occurring.
>
> $$P(X \cap Y) = P(X) \cdot P(Y)$$

Al wonders what independent events would mean in this context. One particular writer meeting a deadline does not seem to affect the likelihood of any other writer meeting the deadline. Thus, the events that writers X and Y meet the deadline are *independent*. Therefore, the multiplication principle can be used. If each of his 10 writers meets the deadline 95% of the time, the probability that all of them meet the deadline is given below.

$$P(\text{going to press on time}) = (0.95)(0.95)(0.95)(0.95)(0.95)(0.95)(0.95)(0.95)(0.95)(0.95)$$
$$= 0.95^{10}$$
$$\approx 0.599$$

Now, for the paper to go to press on time, every one of the 10 writers must meet the deadline. If even one of the ten does not meet the deadline, the printing of the paper is delayed. This calculation shows that *The Lancer* would go to press on time approximately 60% of the time! No wonder it has been going to press late so often. Al wants to change this situation to ensure the paper goes to press on time 90% of the time.

Each writer meeting deadlines 95% of the time seems to be a reasonable expectation. However, having each of the 10 students meet the deadlines 95% of the time results in a big problem for *The Lancer*. To improve this situation, Al believes he can get his editors to meet the deadlines 99% of the time. They have the most experience and are likely to work better under pressure compared to the staff writers who have just joined *The Lancer*'s staff.

Al's idea is to have his five editors meet deadlines 99% of the time, and the less experienced staff writers can continue to meet deadlines 95% of the time. The multiplication principle allows Al to calculate the probability of going to press on time.

$$P(\text{going to press on time}) = (0.99)(0.99)(0.99)(0.99)(0.99)(0.95)(0.95)(0.95)(0.95)(0.95)$$
$$= (0.99)^5 \cdot (0.95)^5$$
$$\approx 0.736$$

Going to press on time about 74% of the time is far better than the 60% it was previously. Nevertheless, he would still not achieve his 90% goal.

Al goes back to the drawing board and believes that he can push his editors to be on time 100% of the time. He feels the editors can handle that level of responsibility. If the five editors meet the deadlines 100% of the time and the staff writers meet the deadlines 95% of the time, the probability of going to press on time is shown below.

$$P(\text{going to press on time}) = (1)(1)(1)(1)(1)(0.95)(0.95)(0.95)(0.95)(0.95)$$
$$= (1)^5 \cdot (0.95)^5$$
$$\approx 0.775$$

Getting the editors to meet their deadlines 100% of the time barely made a difference!

Al sees that the only way to get *The Lancer* to press on time is to push also the staff writers to meet their deadlines at a higher rate. If the editors meet the deadlines 100% of the time, what must the probability be for the staff writers in order for *The Lancer* to not be delayed 90% of the time?

1. If each staff writer meets her/his deadline 96% of the time, would The Lancer go to press on time 90% of the time? If not, what if the staff writers meet their deadlines 97% of the time?

Al begins to wonder whether it is reasonable to have a goal of 90% for *The Lancer* to go to press on time. If the editors meet their deadlines 100% of the time, Al wonders what standard should be required of each staff writer. We can take an algebraic approach to determine this standard.

Let x represent the standard for each staff writer meeting the deadline. The probability of the paper going to press on time at least 90% of the time can be represented by the following inequality. Then we need to solve the inequality for x.

$$P(\text{going to press on time}) \geq 0.90$$
$$(1)(1)(1)(1)(1)(x)(x)(x)(x)(x) \geq 0.90$$
$$1^5 \cdot x^5 \geq 0.90$$
$$x^5 \geq 0.90$$
$$x \geq \sqrt[5]{0.90}$$

Now, take the fifth-root of a number, which is equivalent to raising the number to the $\frac{1}{5}$ power. This can be done with a calculator, so

$$x \geq (0.90)^{\frac{1}{5}}$$
$$x \geq 0.979$$

The solution shows us that for *The Lancer* to go to press on time 90% of the time, each of the five staff writers would have to make deadline almost 98% of the time! Al wonders if this is a realistic goal. In the next chapter, Al will evaluate a change in policy. He will consider going to press on time if one or two writers are late. He will simply leave their columns out of the paper.

1.4.2 Complementary Events

Let's consider the timely publication of *The Lancer* from a slightly different perspective. Instead of focusing on how frequently *The Lancer* goes to press on time, let's consider how often it goes to press late.

Going back to the original case, the editors and staff writers originally each met the individual deadlines 95% of the time. This led to *The Lancer* going to press on time roughly 60% of the time.

2. That being the case, how often did *The Lancer* go to press late?

If they were on time 60% of the time, it makes sense to say that they were late 40% of the time. Mathematically, this happens because going to press on time and going to press late are **complementary events**. Complementary events are both ***mutually exclusive*** and ***collectively exhaustive***. If two events are mutually exclusive, that means they cannot occur at the same time. For example, when rolling a die, the event of getting a three and the event of getting a five are mutually exclusive. However, getting a 3 and getting an odd number are not mutually exclusive. Two events are collectively exhaustive if together they cover all possible outcomes. This implies that one of them must occur. For example, when rolling a die, the event of rolling an odd number and the event of rolling an even number are collectively exhaustive. However, rolling an odd number and rolling a 2 are not collectively exhaustive.

3. Keeping with our die-rolling context, list three pairs of complementary events.

When we consider publishing a student newspaper, one pair of complementary events is publishing on time or not publishing on time (i.e., publishing late). Because an event and its complement include all the possible outcomes of an event, the probability that one or the other of them will occur is equal to one (i.e., one of them is certain to occur). Now, because they cannot both occur (why not?), you can find the probability of an event's complement by subtracting the probability of the original event from one. This is significant because calculating the probability of the complement of an event is sometimes easier than finding the probability of the event directly. Thus, in the case of *The Lancer*,

$$P(\text{publishing on time}) + P(\text{not publishing on time}) = 1$$
$$P(\text{not publishing on time}) = 1 - P(\text{publishing on time})$$

You might think that all of the writers on time and all of the writers not on time are complementary events. All of the writers on time corresponds directly to the event that all ten writers meet their individual deadlines. The complementary event that the paper is late is not equivalent to **all** writers being late. It only takes **one** writer being late for publication of *The Lancer* to be delayed. Thus the event the paper is late includes all of the following possibilities: 1, 2, 3, 4, 5, 6, 7, 8, 9, or 10 writers being late. Calculating directly the probability of the paper being late requires calculating the probability of each of these one through ten possibilities. The calculation of each of these is in itself complex and requires the advanced concepts discussed in the next chapter.

Section 1.5 Worker Absenteeism at BT Auto Industries

Unscheduled absenteeism is the bane of any manager of a manufacturing or service facility. When a worker is absent either some work does not get done, or other workers have to increase their productivity. Alternatively, substitute workers must be called in. A classroom is a good example in the lost value when a teacher is absent and a substitute fills in.

1.5.1 Machine Repair Workers

According to The Centers for Disease Control and Prevention (CDC), the flu season usually ranges from November through March. Historically, the peaks usually occur sometime during a 100 day period from December to early March, with flu outbreaks cascading across the country. The CDC recommends that anyone who contracts the flu stay home to prevent spreading it to others. However, a recent study at the Children's Hospital of Philadelphia documented that most medical professionals at that facility do not follow this advice in their workplace. (JAMA Pediatr. 2015;169(9):815-821)

BT Auto Industries is a small manufacturer based in Detroit, Michigan. One particular piece of equipment is critical to its operation. When the equipment fails and cannot be repaired quickly, production is completely stopped. If the stoppage is long, workers are sent home in the middle of the day. Repair of the equipment requires special training and skills. Currently, two workers in the factory, Alejandro and Bernice have undergone extensive training. They are the only employees who can repair this equipment. David Plante, director of operations, is concerned about the upcoming flu season. There is an estimated 10% probability that a worker will be absent on a particular day due to the flu. His primary concern is the possibility that both Alejandro and Bernice would be absent on the same day.

Let A = the event that Alejandro is not sick and is at work
         B = the event that Bernice is not sick and is at work

The complements of these events are
         A^c = the event that Alejandro is out sick
         B^c = the event that Bernice is out sick

The director of operations quickly estimated the likelihood that both would be absent on the same day. He assumes that the absences are independent events. Thus the multiplication rule can be applied.

$$P(A^c \cap B^c) = P(A^c) \cdot P(B^c)$$
$$= (0.1)(0.1)$$
$$= 0.01$$

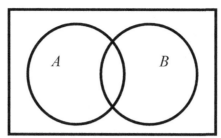

Figure 1.5.1: Venn diagram for Alejandro's and Bernice's presence at work

At first glance, his initial reaction is there is not much to worry about. The probability is only 0.01 or one in 100. However, his production supervisor, Amadeus Wolfe quickly points out that the flu season is 100 days long. Thus, on average there will be one day per season when both Alejandro and Bernice will be absent. In addition, Amadeus noted that some of the more difficult repairs were better handled by two people. In those instances the two working together could repair the equipment more than twice as fast as a single individual. The probability that both are at work is 0.81.

$$P(A \cap B) = P(A) \cdot P(B)$$
$$= (0.9)(0.9)$$
$$= 0.81$$

On average, they will both be at work more than four days out of five. The likelihood that only one will be at work involves considering two mutually exclusive alternatives. Alejandro is at work and Bernice is out with the flu or Alejandro is out with the flu and Bernice is at work. Because these are mutually exclusive events their probabilities can be added.

$$P(\text{exactly one of them is at work}) = P(A \cap B^c) + P(A^c \cap B)$$
$$= (0.9)(0.1) + (0.1)(0.9)$$
$$= 0.18$$

To double-check his calculation, he thinks about the three possible outcomes: both are at work, only one is at work and neither is at work. Since this is the set of all possible outcomes, their probabilities should sum to one.

1. Confirm that the probabilities of these three mutually exclusive events do sum to one.

David asks Amadeus to determine the value of investing $1,000 in a week's worth of training for a third individual named Carmi. To do so, he first calculates the probability that all three would be absent from work on the same day. Again assuming that the three absences are independent events,

$$P(A^c \cap B^c \cap C^c) = P(A^c) \cdot P(B^c) \cdot P(C^c) = (0.1)^3 = 0.001, \quad \text{or} \quad 1 \text{ in } 1,000.$$

Training a third individual reduces the risk of having no one available to fix the machine to an acceptable risk of once in a 1,000 days. Next, Amadeus tries to calculate the likelihood that at least two people are available to work together on repairing the equipment. There are two ways to approach this calculation. The more direct way is for Amadeus to calculate the probability that exactly two are on duty and add it to the probability that all three are at work.

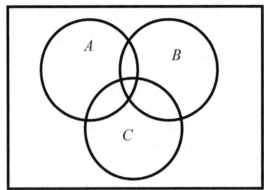

Figure 1.5.2: Venn diagram for Alejandro, Bernice, and Carmi

Let X = number of trained employees on duty.

$$P(X \geq 2) = P(X = 2) + P(X = 3)$$

$$\begin{aligned}
P(X = 2) &= P(A \cap B \cap C^c) + P(A \cap B^c \cap C) + P(A^c \cap B \cap C) \\
&= (0.9)(0.9)(0.1) + (0.9)(0.1)(0.9) + (0.1)(0.9)(0.9) \\
&= 3\left[(0.9)^2 (0.1)\right] \\
&= 3(0.081) \\
&= 0.243
\end{aligned}$$

$$\begin{aligned}
P(X = 3) &= P(A \cap B \cap C) \\
&= (0.9)(0.9)(0.9) \\
&= (0.9)^3 \\
&= 0.729
\end{aligned}$$

$$\begin{aligned}
P(X \geq 2) &= P(X = 2) + P(X = 3) \\
&= 0.243 + 0.729 \\
&= 0.972
\end{aligned}$$

With only two workers, the probability of two being at work was 0.81. However, with the added worker, the probability of at least two workers present increases to 0.972.

Amadeus wondered if it would have been easier to work with the complements to determine this probability.

2. Use the following equation to calculate the same probability.

$$P(X \geq 2) = 1 - P(X \leq 1)$$
$$= 1 - \left[P(X = 0) + P(X = 1) \right]$$

Dionne Trump had listened to the discussions between Amadeus and David. She believed they significantly overestimated the risk because they left out one important factor. They did not include the frequency of the equipment breaking down. If the equipment failed frequently, one or more times per day, then their concerns made sense. However, she was not sure how reliable the equipment was.

3. How would the above analysis change if the equipment broke down on average only once every five days?

4. Would it make sense to increase the maintenance budget by $1,000 so that the frequency of breakdowns was reduced to one in 20 days?

1.5.2 Simulating Day-to-Day Operations

David and Amadeus decide to turn their attention to day-to-day operations. The production floor has 12 workers. In particular, an average of one out of every 10 workers is absent every day during the flu season! In order to maintain their level of productivity, BT Auto Industries hires spare workers to cover the shifts of the employees who have called in sick. Due to tight budget constraints this year, the management wants to hire just enough spare workers to achieve the probability that 95% of the time the absent workers are covered. Anything less and productivity is impacted.

During December, they experienced a number of worker shortages but they were not overly concerned. The demand for their products generally drops during the pre-holiday season. However, they are concerned about the first 10 weeks of the new year when demand picks up. Derek Hall, the plant manager, is interested in determining the number of spare workers to hire on standby. He asked Donald Champ, an industrial engineer, to run a computer simulation of the number of workers who show up each day. The simulation will use random numbers to replicate each worker's attendance during the 10-week peak flu season. They will need to simulate the attendance of each of the 12 workers on each of 50 work days.

An analysis of attendance records for the past five years indicates that an average of one in 10 workers is absent every day during these 10 weeks. They will, therefore, use a 10% absenteeism rate for each of the 12 workers. They generated random integer values between 1 and 10. They assigned a value of 1 to mean that the worker was absent on a given day. An integer between 2 and 10 represented a worker who was present that day.

Table 1.5.1 illustrates the simulation of three days. Each of the random numbers takes on a value between 1 and 10. Table 1.5.2 illustrates how these numbers are then translated into whether or not a worker is absent or present. On Day 1, the number 1 appears only under worker A. There was a total of one absence that day. On Day 2, the number 1 appears only under worker C; thus, there was one worker absent that day. On Day 3, the simulation generated the number 1 under workers E, I, and L. There were three workers absent on Day 3.

Day	\multicolumn{12}{c	}{Workers}	Total Absent each day
	A	B	C	D	E	F	G	H	I	J	K	L	
1	1	6	6	7	2	8	8	4	9	6	9	9	1
2	3	8	1	4	7	9	9	8	8	10	3	10	1
3	6	10	7	6	1	4	2	5	1	3	7	1	3

Table 1.5.1: Simulation of absenteeism on first three days

Day	A	B	C	D	E	F	G	H	I	J	K	L	Total Absent Each day
1	A	P	P	P	P	P	P	P	P	P	P	P	1
2	P	P	A	P	P	P	P	P	P	P	P	P	1
3	P	P	P	P	A	P	P	P	A	P	P	A	3

Table 1.5.2: Translate numbers into Absent (A) or Present (P)

Table 1.5.3 provides the detailed simulated attendance data for each worker on each day for 50 days. It also includes daily totals and individual worker totals for the simulated 10-week period of 50 work days.

If you were to use a calculator to generate data as in Table 1.5.1, you would have to generate the random numbers in stages. The screen of most calculators cannot display more than five numbers at a time. You would then have to record the information in a table. Afterwards, you would identify which values correspond to absent and present. This would be a tedious task. In contrast, Excel can easily simulate 50 days for 12 workers and produce Table 1.5.2 directly. This is demonstrated in section 1.5.3.

Day	Workers												Total absent each day
	A	B	C	D	E	F	G	H	I	J	K	L	
1	1	6	6	7	2	8	8	4	9	6	9	9	1
2	3	8	1	4	7	9	9	8	8	10	3	10	1
3	6	10	7	6	1	4	2	5	1	3	7	1	3
4	10	5	3	7	9	8	9	6	9	9	7	1	1
5	9	10	8	2	5	4	6	5	9	10	3	5	0
6	10	7	5	1	2	2	1	4	6	6	10	3	2
7	8	4	9	4	8	5	3	7	10	10	5	9	0
8	9	5	6	8	2	8	4	10	9	10	7	7	0
9	1	8	2	9	2	3	2	4	6	10	8	7	1
10	6	4	7	4	10	9	2	4	7	10	6	5	0
11	6	4	2	6	7	2	3	5	9	10	4	3	0
12	6	1	8	4	9	6	10	7	9	7	9	8	1
13	9	7	3	9	8	5	9	5	6	6	3	10	0
14	3	8	4	6	7	8	10	9	1	5	1	8	2
15	4	9	5	2	7	5	5	5	8	5	1	5	1
16	5	8	2	10	9	3	7	3	6	2	6	1	1
17	7	10	10	4	4	3	6	3	7	8	7	9	0
18	6	7	9	2	2	6	8	8	5	9	7	2	0
19	9	9	8	3	10	7	7	10	9	1	3	4	1
20	3	9	10	6	7	4	10	8	9	3	9	8	0
21	8	5	8	7	6	1	4	4	8	2	9	7	1
22	9	4	7	6	8	9	2	10	10	7	8	7	0
23	10	2	7	1	4	9	8	1	5	4	9	4	2
24	2	1	5	7	3	1	5	3	4	1	8	4	3
25	5	8	7	9	1	10	9	6	7	5	7	2	1
26	3	2	5	5	2	6	8	5	3	9	9	3	0
27	10	2	8	3	1	10	5	5	9	3	1	2	2
28	4	2	6	9	7	3	7	2	6	7	2	1	1
29	1	2	3	1	10	10	10	4	2	9	2	7	2
30	9	3	5	8	8	1	4	1	10	3	4	5	2
31	3	2	8	1	3	1	5	7	8	1	1	7	4
32	10	3	5	10	6	3	6	2	7	5	1	6	1
33	5	1	9	10	6	10	6	1	10	6	2	10	2
34	3	2	10	7	9	2	8	9	4	4	1	5	1
35	4	9	5	7	1	8	2	4	5	6	4	8	1
36	10	3	9	2	9	3	6	3	1	3	6	6	1
37	5	6	4	4	6	4	7	3	4	2	3	6	0
38	1	6	9	5	7	4	8	10	8	4	7	6	1
39	7	4	1	8	4	6	3	9	4	9	2	6	1
40	8	9	5	10	1	5	1	1	6	8	10	4	3
41	6	9	8	7	7	5	10	1	3	7	5	1	2
42	5	2	3	9	9	6	7	5	3	9	6	9	0
43	9	6	3	10	5	7	2	8	9	1	6	8	1
44	6	5	1	7	4	4	1	7	10	2	9	3	2
45	8	7	4	1	6	7	8	5	1	2	7	9	2
46	8	7	7	3	1	7	5	5	8	3	6	6	1
47	10	1	5	3	3	5	5	7	7	2	4	1	2
48	1	6	6	5	8	7	3	4	8	10	4	6	1
49	2	4	5	2	1	3	10	4	5	4	8	8	1
50	5	1	6	10	10	6	4	6	3	6	9	9	1
Absent	5	5	3	5	7	4	3	5	4	4	6	6	

1 = absent; 2 through 10 = present

Table 1.5.3: Fifty daily results of computer simulation of worker absenteeism

Table 1.5.4 summarizes the frequency distribution of the number of absent workers for the 50-day simulation. The absolute frequency is converted into the relative frequency by dividing each value by 50, the number of simulated workdays. For example, on 13 out of 50 days, there were no workers absent for a relative frequency of 26%. The cumulative relative frequency sums the percentages up to the corresponding value. For example, the cumulative relative frequency of having 3 or fewer absent workers was 98%. This is the sum of the percentages for 0, 1, 2, or 3 absent workers. Use the information in the Table 1.5.3 and 1.5.4 to answer the following questions.

Absent Workers	Absolute Frequency	Relative Frequency	Cumulative Relative Frequency
0	13	26%	26%
1	22	44%	70%
2	11	22%	92%
3	3	6%	98%
4	1	2%	100%
5	0	0%	100%

Table 1.5.4 Frequency distribution of the simulated number of absent workers

5. For what percentage of the 50 simulated days would one spare worker be enough?

6. For what percentage of the 50 simulated days would two spare workers be enough?

7. If the goal is to have enough spare workers 95% of the time, based on this simulation how many spare workers are needed?

8. If the goal is to have enough spare workers 99% of the time, based on this simulation how many spare workers are needed?

9. Can you explain why there are gaps in the actual percentages in questions 5-8?

10. Based on the Table 1.5.3, what is each worker's actual rate of absenteeism? What is the minimum percentage rate and what is the maximum?

11. Why is each worker's actual rate not equal to 10%?

12. What is the overall average absenteeism rate for the 12 workers? Why is this rate closer to the one-in-ten average as compared to many of the individual workers?

13. In setting up the simulation, what assumption have we made about each worker's rate of absenteeism? Do you think the assumption is valid?

1.5.3 Creating Your Own Simulation

Because this is a random process, the results from repeated simulations can be different. To be confident about their spare worker policy, BT Auto's management decides to replicate this 50-day period many times over. They also would like to understand how much variability there might be from year to year in achieving their 95% goal.

Each member of the class will create a spreadsheet to simulate the scenario of a 12-person workforce with a 10% absenteeism rate. The model will show a 50-day period. (If you are reading the text on your own, we have provided an Excel file worksheet with simulated data for an imaginary group of classmates.)

Open Excel and create a table with a column labeled for each worker and a row for each day. Use row "1" to assign an identifying number to each workers. Use column "A" to number the days. Within the table, in the cell for your first worker on the first day, you need to enter a formula whose output will indicate whether that worker was absent on that day. The formula is similar to the "randInt(" command on the graphing calculator. In Excel the formula is "RANDBETWEEN." As you might imagine, this formula will return an integer value between any two integers you input.

For the situation where each worker is absent 10% of the time, we need to think of the absentee rate as a probability. A 10% absenteeism rate means that on any given day a worker may be absent with a probability of one out of 10. Thinking about it this way, we can choose a random number between 1 and 10. Assign an output of "1" to represent a worker being absent. The outputs "2" through "10" represent a worker being present.

To do this, enter an equal sign (=) and the following formula into the appropriate cell.

=RANDBETWEEN(1,10)

Once this formula is entered, an integer between 1 and 10 should appear in the cell. To fill the rest of your spreadsheet, first drag the formula across the row and then down the column to copy it into the remaining cells.

Your spreadsheet should now be populated with integers from 1 to 10 for each of your 12 workers (columns labeled B through M) for 50 days (rows). You will need to prevent these randomly generated numbers from constantly changing. To do this, copy all of the random values using a *copy*, *paste special*, *value* command. You should paste these values into the same cell locations the random numbers originally appear in.

Now we need to determine how many workers were absent each day and how many times each worker was absent during the 50 days. Excel has a formula that enables us to determine this statistic.

The "COUNTIF" formula will search a range of cells and count how many times a particular value occurs. We are concerned with absences, which are indicated by a "1" in a cell. We can give Excel a command to search each row and each column for ones. It will report back how many times the number "1" occurred. You need to include in the formula the range of cells to search and the value it should count.

For example, assume the record of each of the 12 workers on Day 1 is in cells B6 through M6. Below is the appropriate command to count all "1"s.

=COUNTIF(B6:M6, 1)

The resultant count will represent the number of workers absent on day "1". Drag the formula down the column to obtain a value for each of the 50 days.

The resultant count will represent the number of times this worker was absent during the 50 day period. Drag the formula across the row to obtain a value for each worker.

We apply the same type of formula for each column to determine the number of times a specific worker was absent during the 50 days. For example, if the 50 days of data for worker A are in cells B6 through B55, the appropriate command to count all "1"s is below.

=COUNTIF(B6:B55, 1)

Drag the formula across the row to obtain a value for each worker.

14. Compare your row and column totals with those of your neighbors. What do you notice?

15. What could explain the similarities and differences you noted?

16. Can you find anyone else with a simulation spreadsheet that is exactly like yours?

Managers at BT Auto with different responsibilities will look at the data presented in Table 1.5.3 from diverse perspectives. The plant manager is primarily concerned with getting the work done. This official is most interested in how many workers showed up for work. This assumes that all of the workers can do all of the jobs and there is no specialization. The manager will, therefore, focus on the row totals. In a later section, we take the plant manager's perspective and explore the randomness in the number of workers who were absent. This analysis is critical to determining the need and cost of having spare workers to cover for absentees.

The payroll manager is concerned with the details as to who showed up for work each day. He has to submit that information to ensure each individual is paid appropriately. In addition, the worker's absences are recorded against his sick leave allowance. The Human Resource (HR) manager may be interested in the pattern of absences for individual workers. Is a specific worker absent an unusual amount of time? Is another worker always there and perhaps eligible for a bonus? The HR manager will be interested in the column totals for each worker. In the next section we look at the variability in the simulated data from the perspective of each worker's record.

Before proceeding let's explore your gut instinct. Since there is 0.1 probability a worker will be absent on any given day, workers will be absent on average five days out of 50.

17. However, what do you think the chances are that a worker will not miss a single day of the 50 days? Is it one in 10? One in 100? One in 1,000? Or less than one in 1,000?

18. Also, what do you think the chances are that a specific worker will be absent 10 or more days during the flu season?

19. Similarly, what do you think the chances are that all 12 workers will show up for work on a given day?

1.5.4 Individual Worker Absences – Your Individual and Class Pooled Data

The Assistant Vice President of Human Resources, Charlene Fine, is interested in identifying the most reliable workers. These are workers who show up every day no matter what. She is also interested in monitoring those workers who are frequently out sick.

20. From the column totals, identify the minimum and maximum number of days a worker was absent. How many workers were not absent even once during a 50 day period in the spreadsheet in Table 1.5.3? What about the workers in your spreadsheet?

21. Which worker in your simulation had the best attendance record? How much better than the average of five absences was that worker?

22. Which worker had the worst attendance record? How much worse was that worker than the average?

23. Compare your answers with your neighbors. Were the workers with the best and worst attendance records the same for you and your neighbors?

24. What was the least number of days a worker was absent among in your small group? What was the most number of days a worker was absent?

25. Justify whether or not the worker with best attendance in the simulated dataset should be rewarded for good attendance.

You are going to use the class values to develop a frequency distribution table of the number of days a worker was absent during the 50-day time period. You will first do this just for your own data. Then the class will pool all the data to obtain a larger sample. This will be a better estimate of the relative frequency of different values. Then the percentages will be accumulated in order to create a cumulative relative frequency.

Identify the worker(s) with lowest number of absences for the class. This is the class minimum. Identify the worker(s) with highest number of absences for the class. This is the class maximum.

Create a sequence of integer values beginning at the class minimum and ending at the class maximum. Next to each integer value, record the number of times in *your* spreadsheet that each value appeared in the column totals.

26. How many workers in your spreadsheet were absent the class minimum and class maximum number of days?

27. Now you will pool the frequency distribution results for the whole class and then determine the *relative frequency* for the whole class. Find the observed relative frequency for each value for each student and record it in a class pooled table. To start, count the total number of times the minimum value appeared in the pooled data. To obtain the relative frequency, divide this by 12 times the number of students in the class. The 12 represents the number of workers in each spreadsheet. Repeat this for every value in the table to determine its relative frequency.

28. Create another column for cumulative relative frequency. To determine, for example the cumulative relative frequency of five or fewer absences, sum the relative frequency values for zero through five.

29. For class pooled data, what is the relative frequency of a worker being present each and every day during a 50-day period? How does this compare to your gut feel estimate noted in question 17?

30. In the class pooled data, how many times was a worker absent exactly once during a 50-day period? What is its relative frequency?

31. In the class pooled data, how many times was a worker absent one or fewer times during a 50-day period? What is its cumulative relative frequency?

32. In the class pooled data, how many times was a worker absent five or fewer times during a 50-day period? What is its cumulative relative frequency?

33. In the class pooled data, how many times was a worker absent nine or fewer times during a 50-day period? What is its cumulative relative frequency?

34. In the class pooled data, how many times was a worker absent 10 or more times during a 50-day period? What is its relative frequency? How does this relate to the answer to question 33? How does this compare to your gut feel estimate noted in question 18?

1.5.5 Total Number of Workers Absent Each Day

In this section, you will take the plant manager's perspective. The focus will be on the frequency distribution of the number of workers absent each day.

35. In your spreadsheet look at the row totals. What were the minimum and maximum number workers that were absent?

36. How many days experienced no absences? How does this compare to your gut feel estimate noted in question 19?

37. What are the minimum and maximum values for the whole class?

You are going to use the class minimum and maximum values to develop a frequency distribution table of the number of workers absent during each day of the 50-day time period. You will first do this just for your own data. Then the class will pool all the data to obtain larger sample. This is a better estimate of the relative frequency of different values. The relative frequency distribution will then be used to create a cumulative relative frequency distribution.

Create a sequence of integer values beginning at the class minimum and ending at the class maximum. Next to each integer value, record the number of times in your spreadsheet that each value appeared in the row totals. Calculate the relative frequency by dividing by 50.

38. In your row totals, how many days were exactly zero workers absent during the 50-day period?

39. Now tabulate the results for the whole class. How many days was exactly zero workers absent during the 50-day period? For the class as a whole, the total number of simulated days is 50 times the number of students who participated in the activity. If there are 20 students in the class, 1,000 days have been simulated. Divide this total of days with zero absentees by the class total number of simulated days to obtain the relative frequency of observing zero absentees.

40. Repeat this calculation of relative frequency for each value ranging from the class minimum to the class maximum.

41. Also calculate the cumulative relative frequency from the minimum to the maximum observed. The cumulative relative frequency for any value below the minimum is zero. The cumulative relative frequency for any value equal to the maximum or higher is one.

42. How do these relative frequencies compare to the data in Table 1.5.4?

1.5.6 Cost of Spare Workers

The company is considering hiring three steady spare workers to show up each day for work. They are on standby to replace absent workers. To encourage them to show up for work, the company established a compensation policy. These standby workers are paid $55 just for coming to work even if they do no work. However, if the individual actually works, the worker is paid an additional $70, for a total of $125.

Assume three spare workers showed up for work.

43. What is the cost to the company on a day that only one worker was absent. Just one spare worker was needed to work and the other two were sent home?

44. What is the cost to the company on a day that two workers were needed and only one was sent home?

45. On any day when the number of absent workers is three or more, what is the cost for standby workers that day?

46. In general let N be the number of standby workers who actually work. With three workers on regular standby, write an equation for the total cost as a function of N.

In Table 1.5.5 we added a column that includes the cost for standby workers for each day in our simulation reported in Table 1.5.3. It is the next to last column on the right.

47. What is the simulated total 50-day cost for the current policy of three spares on standby? What is the average daily cost?

1.5.7 New Policy: Pay Guarantee

Management is considering paying one worker $100 per day every day no matter what because most days there is at least one worker absent. The other two workers are paid as before, either $55 or $125. If the first worker is not needed for the production line, he is given a special task such as cleaning up some set of files or counting inventory. It is not a high-value job but it keeps him busy. We have added another column to our original table and recorded each day's cost for this new policy.

48. On days with zero absences, what does this new policy cost? Does it cost more or less than before?

49. On days with one absence, how much money does this new policy cost? Is it more or less than before?

50. On days with two absences, how much money does this new policy cost? Three absences?

51. When does the policy cost more money and when does it save money?

52. What is the most common dollar amount under the new policy? Explain why it occurs so often.

53. Based on Table 1.5.5, how many days out of the 50 did the new policy save money?

54. Would you recommend the new policy or the current policy? Explain your answer.

Day	\multicolumn{12}{c	}{Workers}	Absent each day	\multicolumn{2}{c	}{Cost}
	A	B	C	D	E	F	G	H	I	J	K	L		Current	New
1	1	6	6	7	2	8	8	4	9	6	9	9	1	$235	$210
2	3	8	1	4	7	9	9	8	8	10	3	10	1	$235	$210
3	6	10	7	6	1	4	2	5	1	3	7	1	3	$375	$350
4	10	5	3	7	9	8	9	6	9	9	7	1	1	$235	$210
5	9	10	8	2	5	4	6	5	9	10	3	5	0	$165	$210
6	10	7	5	1	2	2	1	4	6	6	10	3	2	$305	$280
7	8	4	9	4	8	5	3	7	10	10	5	9	0	$165	$210
8	9	5	6	8	2	8	4	10	9	10	7	7	0	$165	$210
9	1	8	2	9	2	3	2	4	6	10	8	7	1	$235	$210
10	6	4	7	4	10	9	2	4	7	10	6	5	0	$165	$210
11	6	4	2	6	7	2	3	5	9	10	4	3	0	$165	$210
12	6	1	8	4	9	6	10	7	9	7	9	8	1	$235	$210
13	9	7	3	9	8	5	9	5	6	6	3	10	0	$165	$210
14	3	8	4	6	7	8	10	9	1	5	1	8	2	$305	$280
15	4	9	5	2	7	5	5	5	8	5	1	5	1	$235	$210
16	5	8	2	10	9	3	7	3	6	2	6	1	1	$235	$210
17	7	10	10	4	4	3	6	3	7	8	7	9	0	$165	$210
18	6	7	9	2	2	6	8	8	5	9	7	2	0	$165	$210
19	9	9	8	3	10	7	7	10	9	1	3	4	1	$235	$210
20	3	9	10	6	7	4	10	8	9	3	9	8	0	$165	$210
21	8	5	8	7	6	1	4	4	8	2	9	7	1	$235	$210
22	9	4	7	6	8	9	2	10	10	7	8	7	0	$165	$210
23	10	2	7	1	4	9	8	1	5	4	9	4	2	$305	$280
24	2	1	5	7	3	1	5	3	4	1	8	4	3	$375	$350
25	5	8	7	9	1	10	9	6	7	5	7	2	1	$235	$210
26	3	2	5	5	2	6	8	5	3	9	9	3	0	$165	$210
27	10	2	8	3	1	10	5	5	9	3	1	2	2	$305	$280
28	4	2	6	9	7	3	7	2	6	7	2	1	1	$235	$210
29	1	2	3	1	10	10	10	4	2	9	2	7	2	$305	$280
30	9	3	5	8	8	1	4	1	10	3	4	5	2	$305	$280
31	3	2	8	1	3	1	5	7	8	1	1	7	4	$445	$420
32	10	3	5	10	6	3	6	2	7	5	1	6	1	$235	$210
33	5	1	9	10	6	10	6	1	10	6	2	10	2	$305	$280
34	3	2	10	7	9	2	8	9	4	4	1	5	1	$235	$210
35	4	9	5	7	1	8	2	4	5	6	4	8	1	$235	$210
36	10	3	9	2	9	3	6	3	1	3	6	6	1	$235	$210
37	5	6	4	4	6	4	7	3	4	2	3	6	0	$165	$210
38	1	6	9	5	7	4	8	10	8	4	7	6	1	$235	$210
39	7	4	1	8	4	6	3	9	4	9	2	6	1	$235	$210
40	8	9	5	10	1	5	1	1	6	8	10	4	3	$375	$350
41	6	9	8	7	7	5	10	1	3	7	5	1	2	$305	$280
42	5	2	3	9	9	6	7	5	3	9	6	9	0	$165	$210
43	9	6	3	10	5	7	2	8	9	1	6	8	1	$235	$210
44	6	5	1	7	4	4	1	7	10	2	9	3	2	$305	$280
45	8	7	4	1	6	7	8	5	1	2	7	9	2	$305	$280
46	8	7	7	3	1	7	5	5	8	3	6	6	1	$235	$210
47	10	1	5	3	3	5	5	7	7	2	4	1	2	$305	$280
48	1	6	6	5	8	7	3	4	8	10	4	6	1	$235	$210
49	2	4	5	2	1	3	10	4	5	4	8	8	1	$235	$210
50	5	1	6	10	10	6	4	6	3	6	9	9	1	$235	$210
Absent	5	5	3	5	7	4	3	5	4	4	6	6		$12,240	$11,900

Table 1.5.5: Workers' absences and standby worker costs

Chapter 1 (Basic Probability) Homework Questions

Complement
1. Two Complementary events
    a. Describe complementary events with regard to the outcome of a baseball game.
    b. Be careful to describe complementary events with regard to the outcome of a football or soccer game. How is this answer different than the answer to a?
    c. Describe complementary events with regard to performance on a test.
    d. Describe complementary events with regard to the launch of the Space Shuttle.

2. Provide examples of events that are mutually exclusive but are not complementary in the context of:
    a. The outcome of a soccer game or college football game.
    b. Performance on a test
    c. Timeliness of completing a project

Simulation
3. Simulate by hand flip coin – Two coins
    a. Flip two coins. List all of the possible outcomes of the experiment.
    b. In the coin flip experiment if you only record the number of heads, list all of the possible outcomes of the experiment?
    c. Provide an example of complementary events with regard to the outcome of flipping two coins
    d. Flip two coins 10 times. Record your results into the Table 1 below. (Number of Heads equals 0 means experiment resulted in all Tails.)

Number of Heads	Experiment										Absolute Frequency	Relative Frequency	Cumulative Relative Frequency
	1	2	3	4	5	6	7	8	9	10			
0													
1													
2													

Table 1: Record 2 coin flip experiment

   e. How many of the times did you get one Head and one Tail in 10 experiments? Write your answer into the corresponding row under the absolute frequency column.
   f. Write the frequency of other possibilities into the table.
   g. Calculate the relative frequency of each possibility and compare relative frequency of each outcome with two of your friends.
   h. Calculate the cumulative relative frequency of each possibility and compare the cumulative relative frequency of each outcome with two of your friends.

Chapter 1   Basic Probability and Randomness

4. Simulate by calculator flip coin – Two coins
   a. Go to the APPS menu of your calculator, press 0 for probability simulation, and then press 1 to toss coins. Press on the SET button: change the number of trial set to 50, change the number of coins to 2, and press the OK button. Now, press on the TOSS button. You will see a graph with bars showing the frequency of each possible outcome. What do you observe in the graph?
   b. Now, go to the DATA menu and count the frequency of each outcome.
   c. What are the relative frequencies of each outcome? Compare your results with your friends.

5. Calculate Probabilities – Use Multiplication Rule (Two coins)
   a. Calculate the probability of two heads when two coins are flipped
   b. Calculate the probability of zero heads when two coins are flipped
   c. Use complement to calculate the probability of exactly one head when two coins are flipped
   d. Compare the theoretical probabilities with the observed relative frequencies you found in questions 3 and 4.

6. Simulate by hand flip coin – Three coins
   a. Flip three coins. List all of the possible outcomes of the experiment.
   b. If you only record the number of heads, list all of the possible outcomes of the experiment?
   c. In the three coin flip experiment, you can have either all (three) Heads or less than three (zero, one, or two) Heads. The probability of having three Heads or less than three Heads add up to one. What are the other complementary events with regard to the outcome of flipping three coins?
   d. In the three coin flip experiment, one Head and two Tails and three Heads and zero Tails are mutually exclusive events and they are not complementary. Give an example of two mutually exclusive events that are not complementary.
   e. Flip three coins 10 times. After each flip, record the number of heads. Next, in Table 2 below record a 1 corresponding to the number of heads observed on that experiment.

Number of Heads	Experiment										Absolute Frequency	Relative Frequency	Cumulative Relative Frequency
	1	2	3	4	5	6	7	8	9	10			
0													
1													
2													
3													

Table 2: Record 3 coin flip experiment

f. Count and record the number of times the outcome is all Heads and its complement. Calculate the relative frequency. Do relative frequencies add to one?
g. What is the complementary event to observing all Heads? Calculate its relative frequency. How does the answer relate to the value obtained in the previous question?
h. Determine the total frequency of for each value: 0, 1, 2, and 3.
i. Calculate relative frequency of each value.
j. Calculate the cumulative relative frequency. What is the relative frequency of obtaining 2 or fewer heads?
k. From the table above look at the relative frequencies of each outcome. Do they add up to 1?
l. In part d above, you created two mutually exclusive events. What are their relative frequencies? Do these relative frequencies add up to one?
m. The probability of getting only one Head or only one Tail in flipping three fair coins must be equal. Is your result a representation of this statement?

7. Simulate by calculator flip coin – Three coins
   a. Go to the APPS menu of your calculator again. Press 0 for probability simulation, and then press 1 to toss coins. Press on the SET button: change the number of trial set to 50, change the number of coins to 3, and press the OK button. Now, press on the TOSS button. You will see a graph with bars showing the frequency of each possible outcome. What do you observe in the graph?
   b. Now, go to the DATA menu and count the frequency of each outcome.
   c. What are the relative frequencies of each outcome? Compare your results with your friends.
   d. What are the cumulative relative frequencies of each outcome? Compare your results with your friends.

8. Calculate Probabilities – Use Multiplication Rule (Three coins)
   a. Calculate the probability of three heads when three coins are flipped
   b. Calculate the probability of zero heads when three coins are flipped
   c. Explain why the probability of getting one head and two tails should be the same as the probability of getting two heads and one tail.
   d. Use Complement and the answer to part c to calculate the probability of exactly one head. Exactly two heads.
   e. Compare the theoretical probabilities with the observed relative frequencies you found in questions 6 and 7.

9. Physical simulation – Pick one with replacement
   a. Take two red and four blue M&M's® and put them into a bag. Then, blindly withdraw an M&M®. Record your outcome as R or B. Put the M&M® that you picked back into the bag. Perform this experiment six times. How many times did you pick a red M&M® in six trials?
   b. What is the relative frequency of picking R? What is the relative frequency of picking B?
   c. Do the relative frequencies you found in part (b) make sense? Why or why not?

d. Compare the relative frequency of picking an R to 1/3.
e. Record your results from part a in the Table 3 below. Repeat part (a) nine more times. Namely you will blindly withdraw an M&M® and put it back into the bag. Record in Table 3 the result of each withdrawal. Each set of six selections with replacement represents a single trial. Repeat this set of six withdrawals with replacement a total of 10 times.

Trial	Withdrawal						# of Reds	Relative Frequency of Reds
	1	2	3	4	5	6		
1								
2								
3								
4								
5								
6								
7								
8								
9								
10								

Table 3: Simulate selection of M&M's®

f. How many trials recorded a relative frequency of exactly 1/3?
g. Pool your results from all 10 trials. There were a total of 60 selections. Is the relative frequency of picking Reds close to 1/3? Explain why the pooled data should be closer to 1/3 than the individual row recorded relative frequency.

10. Physical simulation – Pick two with replacement and record the values.
    a. You have two red and four blue M&M's® in the bag. Blindly withdraw an M&M®, look at its color. Put it back into the bag. Withdraw a second M&M®, look at its color and put it back into the bag. Record in Table 4 what you observed for the pair of withdrawals.
    b. Repeat this experiment 20 times. Record your results in Table 4.

Outcomes	Trial																				Frequency	Relative Frequency
	1	2	3	4	5	6	7	8	9	10	11	12	13	14	15	16	17	18	19	20		
RR																						
RB																						
BR																						
BB																						

Table 4: Simulate selection of M&M's® with replacement – record actual

c. Which was the most frequent outcome? Which was the least frequent outcome?
d. Now, record in Table 5 your results as the "number" of reds. Both RB and BR equal one red.

Number of Reds	Trial																				Frequency	Relative Frequency
	1	2	3	4	5	6	7	8	9	10	11	12	13	14	15	16	17	18	19	20		
0																						
1																						
2																						

Table 5: Simulate selection of M&M's® with replacement – record number of reds

e. Calculate theoretical probabilities of having zero, one, or two reds in the two M&M® withdrawal experiment.
f. Compare the relative frequencies you observed in part (d) with the theoretical probabilities.

11. Using MS Excel Random Generator
    a. Repeat the experiment "pick two with replacement" 50 times using the Randbetween function of Excel. Set up the Excel sheet as in the Figure 1. Write the formula in cell B2 as in the Formula Bar and copy it to other cells.

The formula, IF(RANDBETWEEN(1,6)<=2,"R","B") generates a random number between one and six. If the number is two or less, it records an R in the cell. Otherwise it records a B. Copy this function into cell C2. Then copy the values in cells B2 and C2 all the way down to cells B51 and C51. To stabilize the random numbers, use the copy and past values commands. (If you do not do this, the random numbers will constantly change.) In column D you can use the CountIf command to count the number of Rs.

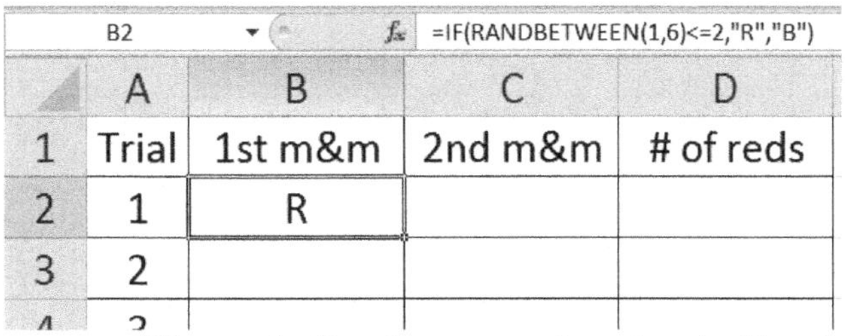

	A	B	C	D
1	Trial	1st m&m	2nd m&m	# of reds
2	1	R		
3	2			
4	3			

B2 =IF(RANDBETWEEN(1,6)<=2,"R","B")

Figure 1: Randbetween function in Excel

b. Record in Table 6 the number of times you observed zero, one, or two Rs.

Number of Rs	Frequency	Relative Frequency
0		
1		
2		

Table 6: Simulate 50 times the selection of M&M's® with replacement – record number of reds

c. Compare the relative frequencies you observed in part (b) with the theoretical probabilities determined in question 10e. Are the relative frequencies closer to the theoretical probabilities when compared to physical simulation that was repeated only 20 times?

12. Multiple Choice Exam
    A student has to take an exam, but he has not studied at all. Therefore, he has no idea about the answers. The exam is a multiple choice exam with 10 questions; each has four choices with only one correct answer. The passing grade for this exam is six correct answers. The student has decided to take his chances. Table 7 contains simulated data for 30 repetitions of a 10-question exam in which a student randomly guesses the correct answer. (C= Correct and I = Incorrect) Answer the questions below according to the data in Table 7.

	Questions										No. of
Trial #	1	2	3	4	5	6	7	8	9	10	Cs
1	C	C	I	I	I	C	I	C	C	C	
2	I	C	I	C	I	I	C	I	I	I	
3	I	I	C	I	I	I	I	I	C	C	
4	C	I	I	I	C	I	I	I	I	I	
5	I	I	C	C	I	I	C	I	C	I	
6	I	I	I	I	I	C	I	I	I	I	
7	I	I	C	C	C	C	I	C	C	I	
8	I	I	C	I	I	I	I	I	I	I	
9	I	I	C	I	I	I	I	I	I	I	
10	I	I	I	I	I	I	I	I	I	I	
11	C	C	C	I	C	I	I	I	C	C	
12	I	C	C	I	I	I	I	C	I	I	
13	I	I	I	C	I	I	I	C	C	I	
14	I	I	I	I	I	I	I	I	I	I	
15	I	C	I	C	I	I	C	I	I	I	
16	I	I	C	I	I	I	I	I	C	I	
17	I	I	I	C	C	I	C	I	I	I	
18	I	C	I	C	I	I	C	I	I	I	
19	C	I	I	C	C	I	I	I	C	I	
20	C	I	I	I	I	I	I	I	C	I	
21	C	I	C	I	I	C	I	C	I	I	
22	C	C	I	I	I	I	I	I	C	I	
23	I	I	I	I	I	I	C	I	C	I	
24	I	I	I	I	I	C	I	I	I	I	
25	I	I	I	I	I	I	C	I	C	I	
26	I	I	I	I	C	C	I	I	I	C	
27	I	I	C	I	C	C	I	I	I	I	
28	I	I	C	C	I	I	C	I	I	C	
29	I	C	I	C	I	I	I	C	I	I	
30	I	I	C	I	C	I	C	I	C	I	
No. of Cs											
Proportion											

Table 7: Multiple choice exam with random guesses

a. Record the number of correct answers in each trial in the last column. How many of the trials did the student fail to score a six or higher?
b. Record the number of correct answers for each question in the last row. Calculate each proportion. Look at the proportion of correct answers to each of the questions. Is it exactly 0.25?

c. Using the last column of Table 7, record in Table 8 the frequency of the number of correct answers on each 10 question exam.

No. of Cs	Frequency	Relative Frequency	Cumulative Relative Frequency
0			
1			
2			
3			
4			
5			
6			
7			
8			
9			
10			

Table 8: Frequency distribution of the number of correct answers

d. What is the most frequent score on the exam? Is this number surprising?
e. What is the observed relative frequency that the student has six correct answers?
f. What is the relative frequency of scoring five or fewer correct answers?
g. What is the relative frequency of scoring six or more correct answers?
h. What is the observed relative frequency that the student has zero correct answers?
i. Calculate the theoretical probability of getting zero correct answers using the Multiplication Rule. (In Chapter 3, you will learn the formula used to calculate all of the theoretical probabilities for each of the possible outcomes.)

13. Simulate a Multiple Choice Exam
    a. Simulate his performance in MS Excel; make 30 simulation runs. Set the Excel sheet as in Figure 2. Write the formula in cell B3 as it appears in the Formula Bar. Copy it to other cells. The IF command in Excel states that if the random number is equal to 1, Excel will record a value of "C" for correct in the cell. Otherwise it records and "I" for incorrect.

B3				f_x	=IF(RANDBETWEEN(1,4)=1,"C","I")						
	A	B	C	D	E	F	G	H	I	J	K	L
1						Question Number						
2	Trial	1	2	3	4	5	6	7	8	9	10	Number of Cs
3	1	C										
4	2											
5	3											
6	4											

Figure 2: Simulate multiple choice exam in Excel

b. Why is there an "= 1" in the statement?
c. He will pass the exam if he answers at least six of the questions correctly. According to spreadsheet experiment, how many trials resulted in failure?
d. In any of the trials, did he answer all the questions correctly?

e. Record in Table 9 the number of correct answers and identify which is the most frequent score.

No. of Cs	Frequency	Relative Frequency	Cumulative Relative Frequency
0			
1			
2			
3			
4			
5			
6			
7			
8			
9			
10			

Table 9: Simulated data - frequency distribution of the number of correct answers

14. Refer to Section 1.5.1 which discusses absenteeism of machine repair workers. David Plante, director of operations at BT Auto Industries, wondered what would happen if there were an unusually serious flu outbreak. In that case absenteeism could be as high as 15%. When the rate was 10%, David felt that it was necessary to train a 3rd worker and that would address his concerns.
    a. Assume absenteeism is 15% and there are three workers trained to repair machines. What proportion of days would three trained workers be enough to ensure that there was always at least one machine repair worker on duty? (Use basic concepts of probability to answer this question.)
    b. Assume you want to simulate absenteeism in a serious flu outbreak. What is the range of values you will use in your random number generator?
    c. What values correspond to a worker being absent with 0.15 probability?
    d. Now simulate 50 days of three worker attendance using a 15% absentee probability. Record your results in a table similar to Table 10.

Day	Worker			Total at Work
	A	B	C	
1				
2				
3				
.				
.				
.				
50				

Table 10: Trained worker attendance during a serious flu outbreak

e. How many times was there at least one trained worker at work? What is the relative frequency? Compare this result to the theoretical probability calculated in part a.
f. Recall that there was significant value in having two workers to jointly repair a machine. How many times were there at least two trained workers at work? What is the relative frequency?
g. Would three trained workers adequately meet the plant's needs even during a serious outbreak of flu? Explain your answer.

Expanded Homework Questions

I. X-Press School Newspaper

X-Press is a monthly school newspaper which started in September 2008. It has three writers and three editors. According to the records from last year, it is estimated that the probability that each student will meet the deadline is 90%. Even though, this year has started out well for all the writers, the paper was delayed twice in six months due to lateness by an editor. Table 11 shows the data for the six months of 2009.

Month	Editors			Writers			Paper Late?
	Tom	Bob	May	Ann	Lee	Eric	
1	on time	on time	on time	on time	on time	on time	No
2	on time	on time	on time	on time	on time	on time	No
3	on time	on time	late	on time	on time	on time	Yes
4	on time	on time	on time	on time	on time	on time	No
5	on time	on time	on time	on time	on time	on time	No
6	on time	late	late	on time	on time	on time	Yes
Number of late times per student	0	1	2	0	0	0	

Table 11: September 2008-February 2009 performance data

1. The data in Table 11 shows that the student writers' on time performance was 100% in the last six months.
   a. Is this strong evidence of an improvement of the timeliness of the student writers?
   b. The two times late in the last six months was due to two of the editors, Bob, and May. Looking at the data, can we say that the editors have some problems?
   c. May was late twice over the six months period. Do you think May has some problems?

Actually, the data in Table 11 was randomly simulated. It is formed by generating random numbers for all the editors and writers as if all of them having a likelihood of 0.9 of meeting each deadline. It is just pure randomness that all the writers in this simulation met the deadline 100% of the time. Therefore, these results provide no evidence that they will show the same performance in the next month.

   d. If you assume that each writer and editor is 90% reliable, what is the probability of all of them meeting the deadline in any month?
   e. What is the probability of the paper being late in any month?

2. Simulate the performance of the newspaper over the 6 months yourself. Create a table as Table 11 above in Excel. Write **=IF(RANDBETWEEN(1,10)=1,"late","on time")** in B3 and copy it to other cells until G8. In cell H3, write **=IF(COUNTIF(B3:G3,"late")>0,"Yes","No")** and copy it until H8. To count the number of lateness of an individual student write **=COUNTIF(B3:B8,"late")** in cell B9 and copy it until G9.

	B3		f_x	=IF(RANDBETWEEN(1,10)=1,"late","on time")			
	A	B	C	D	E	F	G	H
1		Editors			Writers			
2	Months	Tom	Bob	May	Ann	Lee	Eric	Late?
3	1	on time						
4	2							
5	3							
6	4							
7	5							
8	6							
9	Number of late times/student							

Figure 3: Simulation of X-Press Newspaper

a. Is there any editor or writer, whose performance is similar to May's in your experiment? Namely was any editor or writer late twice over the six-month period?
b. Make nine copies of rows one through nine in other rows of the spreadsheet. In other words, simulate the performance of the newspaper over the six months a total of 10 times.
c. In how many of the 10 runs, was a specific editor or writer late twice over the six month period?
d. How many times out of the 10 replications was the newspaper published on time for each and every one of the six months?
e. The ten runs represent simulating 60 months. What was the overall percentage of months the paper was late?
f. Record in Table 12 the results of your 10 simulation runs. Calculate the relative frequencies.

	Number of times late in six months	Relative Frequency	Cumulative Relative Frequency
0			
1			
2			
3			
4			
5			
6			

Table 12: Frequency distribution of number of times late in six months

g. Calculate the cumulative relative frequencies of the number of times late.
h. What is the relative frequency of the student paper being late two or more times over the six-month period?

3. The newspaper has changed its policy. Now, it will publish the paper on time if just one of the students has not submitted his work on time. The newspaper will still be late if two or more are late. Use the previously simulated data to evaluate this new policy.
    a. Make a frequency table similar to Table 12 which shows the number of times the paper is late in 10 simulation runs. Calculate the relative frequencies and cumulative relative frequencies.
    b. How much of a change was there in the relative frequency of never being late during a six-month period when compared to the policy used in question 2?

II. High School Hockey – Five Game Series

1. The Red Run School District has two hockey teams, the Quacks and the Tops. They are going to play a best-of-five series to determine the champion this year. The series ends when one team wins three games. We believe that both teams are equally likely to win in each game.
    a. Take a piece of paper, and cut it into two equal pieces. Write "Quacks" on one piece, and "Tops" on the other piece. Fold the papers, shuffle them, and pick one piece. Record what you have observed. Refold the piece, reshuffle and select again. Repeat this until you record three of the same team's name. The total number of repetitions must not exceed five. It could be as few as three.
    b. How many times did you fold-shuffle-pick until you collected three of the same team's name?
    c. Repeat the experiment in part (a) a total of 10 times. Record the results in a table similar to Table 13.

Trials	1	2	3	4	5	Number of Games
1	Q	T	T	T		4
2	Q	Q	Q			3
3	T	T	Q	Q	T	5
4						

Table 13: Physical simulation of a five game series

   d. How many trials resulted in the order of "Quacks – Quacks – Tops – Tops – Tops"?
   e. What is the probability of this specific sequence, "Quacks – Quacks – Tops – Tops – Tops"? Compare this probability with the relative frequency in your simulation. Are you surprised by the result?
   f. Tabulate the data you created and complete Table 14. This summarizes the relative frequency of the number of games played until a winner was determined.

Games Played	Number of Trials	Relative Frequency	Cumulative Relative Frequency
3			
4			
5			

Table 14: Physical simulation of a five game series - relative frequency

g. What proportion of times did the series last only three games?
h. What is the theoretical probability the series would last only three games? (Remember that there are two distinct ways this can happen.) Compare this probability to the observed relative frequency.

2. Now, let's simulate the series by flipping a coin. Let heads be the Quacks and tails be the Tops. Repeat flipping the coin until you get either three heads or three tails. The total number of flips must not exceed five. This will simulate one five-game series.
    a. Do the same thing 20 times to complete the simulation runs and record the results in a table similar to Table 13.
    b. How many times did Tops win the series in the 20 simulated experiments? Compare your results with your friends.
    c. What proportion of times did the series last only three games?
    d. What other physical experiments can you use to simulate the series? How?

3. Suppose 3 games have already been played, and the Quacks have won two out of three games.
    a. Assume that both teams are equally likely to win in each additional game. What is the probability that the Tops will come back and win the series?
    b. Historical data suggests that a team winning two out of the first three games will win the series 85% of the time. There is only a 15% chance for the other team to win. This percentage is smaller than the value found in part a. According to this information, the Quacks and the Tops are not equally likely to win in each game. Assume the probability that the Tops will win two games in a row is only 0.15. What is the probability of Tops winning an individual game?

4. You will simulate the fourth game using the RAND (Math → Prb → 1: Rand) function of your calculator. If the random number is less than or equal to the probability of the Tops winning that was found in 3b, then the Tops win the game and the series continues. Otherwise the Quacks win the game and the series ends.
    a. Simulate the fourth game. Who won the game?
    b. Simulate the fourth and the fifth games 20 times. What percentage of time did the series end with the fourth game?
    c. Explain why you cannot flip a coin to simulate the fourth game?

5. Simulate the fourth and the fifth games 100 times using MS Excel. Write the formulation in cell B2 and drag it to the bottom and to the right until you cover all the games and the simulation runs. The letter x corresponds the probability of the Tops winning that you have found earlier.

	A	B	C	D
1	Run Number	4th game	5th game	
2	1	=IF(RAND()<=x,"Tops","Quacks")		
3	2			
4	3			

Figure 4: Simulate fourth and fifth games of hockey series

a. How many times did Quacks win the series in four games?
b. How many times did Quacks win the series in five games?
c. How many times did Tops win the series?

III. 2010 Bradley Cup – Seven Game Series

1. The Finals of 2010 Bradley Cup Basketball Playoffs have begun. The Raleigh Elks and Lansing Moose are playing a best-of-seven series for the championship. Four games have already been played. The Elks won three out of the first four games and the Moose won only one game. Assuming that the Moose and the Elks are equally likely to win in each game, answer the three questions below.
    a. What is the probability that the Moose will win the series?
    b. If the Moose wins the fifth game, what is the probability that the Moose will win the series?
    c. What is the probability that the Elks will win the series if the Moose wins the fifth game?

    Historical data indicates that a team winning three out of the first four games won the series 95% of the time. This data suggests that the two teams are not equally likely to win each game.
    d. Explain why the team that is ahead three to one may have a higher than 50% chance of winning each game.
    e. Perhaps the chance of winning in each game is only 40% for the Moose. What would be the probability that the Moose will win the series?
    f. Based on historical data, the chances of a team like Moose winning the series is only 0.05. What value raised to third power yields 0.05? What does this number represent?

2. Simulate the fifth, sixth and seventh games 100 times using MS Excel. Use the value you found in part (f) above rounded to three decimal places to run your simulation.
    a. How many times did Elks win the series in five games?
    b. How many times did Elks win the series in six games?
    c. How many times did Moose win the series?

IV. Airline Overbooking

1. Profits and Break-Even Analysis

Great Lakes Airlines (GLA) operates out of Lansing, MI (LAN) and serves 20 cities in seven states. They want to conduct a profit break-even analysis on Flight 425 which flies from Lansing (LAN) to in Milwaukee, WI (MKE). The average one-way airfare price is $300. The basic operating cost of the plane is $6,000 for each flight with an additional cost of $20 per customer. Flight 425 has 30-passenger seat capacity. Customers who cancel their flights or don't show up at the gate will pay GLA $ 75 but get back the rest of the ticket price.

a. What is the revenue if the number of reserved passengers is 20 and all passengers show up?
b. What is the cost if the number of reservations is 20 and all passengers show up? What is the profit or loss with 20 passengers?
c. The number of passengers where the cost equals revenue is called the break-even point. According to this, what is the break-even point for this case in which all reservations ow up?

According to GLA flight records, 15% of the people who make reservation either cancel or just don't show up.

d. In Table 15 below, five customers out of 30 did not show up, so what is the profitability? (Remember to include the penalty cost for not showing up.)

Customer Number	Show-up?	Customer Number	Show-up?
1	Yes	17	Yes
2	No	18	No
3	Yes	19	Yes
4	Yes	20	Yes
5	Yes	21	Yes
6	Yes	22	No
7	Yes	23	Yes
8	Yes	24	Yes
9	Yes	25	Yes
10	Yes	26	Yes
11	Yes	27	Yes
12	Yes	28	No
13	Yes	29	No
14	Yes	30	Yes
15	Yes	Number of No-Shows	5
16	Yes		

Table 15: Customers showing up out of total of 30 reservations

e. Write a mathematical expression to calculate GLA's net profit as a function of the number of reservations and the number of no shows. Be careful to define all of your variables.
f. Assume that 30 customers have reserved tickets for a flight. Simulate the showing up behavior of each customer in Excel Worksheet. Use Figure 5 as your reference. Then calculate the net profit GLA earned on the flight. Use RANDBETWEEN(1,100) function to generate numbers between 1 and 100. If this random number is less than or equal to 15, it means the person did not show up. Count the number of no-shows using the function COUNTIF(B2:B31,"No").

	A	B	C	D
1	customer #	Show-up?		
2	1	Yes		
3	2	Yes		
4	3	Yes		
5	=IF(RANDBETWEEN(1,100)<=15,"No",			
6		"Yes")		
7	6	Yes		
8	7	Yes		
9	8	Yes		
10	9	Yes		
11	10	No		
12	11	Yes		

Figure 5: Simulation of customers showing up at the gate

    g. Is it possible to fill all the seats without booking a maximum of 30 passengers? What is the probability of that happening?

    h. If GLA has 250 flights per year, on average how many times will all 30 people show up?

2. Overbooking and Bumping: 31 Reservations Allowed

Overbooking means that more tickets are sold (or booked) than there are seats on the plane. Overbooking is a common practice in the airline industry and it is legal. Sometimes people cancel their flights or just don't show up at the gate. This results in empty seats on the planes and loss opportunity for more revenue. By overbooking flights, the likelihood of empty seats because of those cancellations and no-shows is reduced.

If the number of ticketholders who arrive at the gate is more than the flight capacity, then some of the passengers will be bumped to other flights. They will receive some compensation for the disruption. They generally receive another ticket for the next available flight and compensation in the form of cash or a voucher for future travel. If the delay involves an overnight stay, they will be placed in a hotel free of charge. Compensation will vary depending upon whether the passenger volunteers to be bumped or is involuntarily bumped.

GLA wants to determine a good overbooking strategy to reduce the lost revenue due to cancellations and no shows on fully booked flights. They will apply the initial overbooking strategy on Flight 425. If a customer is denied boarding in case of overbooking, GLA will bump the customer to another flight. GLA will $250 to compensate that customer for the inconvenience. GLA starts its overbooking strategy with 31 reservations on each flight.

    a. Determine GLA's net profit on this flight if all 31 customers show up. (Do not count the 31st ticket as revenue for this flight. The bumped customer will use his ticket on another flight.)

b. Determine GLA's net profit if only 30 customers show up. (Remember a no-show is charged only a $75 penalty and not the full $300 price of the ticket.)
c. Determine GLA's net profit if only 29 customers show up.
d. Write a mathematical expression to calculate GLA's net profit for the case of overbooking as a function of: the number of reservations, the number of no shows and the number of people bumped. Be careful to define all of your variables.
e. Table 16 shows the frequency of the number of customers who did not show up over 100 days when GLA allows 31 reservations and overbooks one customer consistently. Fill in the blank cells under the number bumped and the profit/day column. Multiply the number of days by profit/day. Write the solution in the fifth column. Take the average of the fifth column to find the average profit per day.

Number of No-Shows	Number Bumped	Profit/Day	Number of Days	Profit/Day × Number of Days
0			1	
1			1	
2			8	
3			17	
4			23	
5			15	
6			15	
7			7	
8			6	
9			5	
10			2	
			Average Profit/Day	___

Table 16: Frequency of no shows for 100 days with 31 bookings allowed

f. What number of no-shows was the most profitable? Why?

3. Overbooking and Bumping: 32 Reservations Allowed

a. Table 17 shows the frequency of number of no-shows over 100 days when GLA allows 32 reservations. Fill in the blank cells under the number bumped column. Calculate the relative frequency for each number of no-shows and write it to the corresponding row in the fourth column. Write the cumulative relative frequencies in the fifth column.

Number of No-Shows	Number Bumped	Number of Days	Relative Frequency	Cumulative Relative Frequency
0		1		
1		1		
2		8		
3		17		
4		23		
5		15		
6		15		
7		7		
8		6		
9		5		
10		2		

Table 17: Frequency table for 100 days

Now, answer the questions below using the fifth column of Table 17.
  b. What percent of the time did all 32 customers appear at the gate?
  c. What percent of the time was at least one customer bumped?

  d. What percent of the time did more than four customers not show up for their flight?
  e. Complete Table 18 to determine the average profit per day..

Number of No-Shows	Number Bumped	Profit/Day	Number of Days	Profit/Day × Number of Days
0			1	
1			1	
2			8	
3			17	
4			23	
5			15	
6			15	
7			7	
8			6	
9			5	
10			2	
			Average Profit/Day	___

Table 18: Frequency of no shows for 100 days with 32 bookings allowed

4. Optimal Overbooking Policy

Table 19 shows the 100-day simulation results of Flight 425 with different overbooking policies (35, 36, and 37 reservations). Each policy was simulated independently. For example, according to this table, when GLA allowed 35 customers to reserve for Flight 425, three customers did not show up eight times out of 100 days. In this case, two customers of the 32 customers who did show up have to be bumped to other flights. Use the data in Table 19 to answer the following questions.

Number of No-Shows	Frequency of No-Shows for Different Policies		
	35	36	37
0	2	1	0
1	2	0	3
2	7	5	2
3	8	2	9
4	16	17	14
5	21	21	14
6	16	21	17
7	12	16	16
8	8	9	14
9	6	3	7
10	0	3	2
11	2	0	2
12	0	2	0
13	0	0	0
14	0	0	0
15	0	0	0

Table 19: No-Shows history of Flight 425 for different overbooking policies

a. Fill in the blank cells in Table 20 using the data from Table 19.

	Overbooking Policy		
	35	36	37
Probability of no bumping			
Probability of bumping 2 or more			
Average number of no-shows			
Average number of bumped passengers			
Average number of flying passengers			

Table 20: Evaluation of overbooking policies – 35, 36, and 37

b. Create a table similar to Table 18 for each of these three overbooking policies.
c. Determine the average profit for each policy? Which policy is the best?
d. How many random numbers would you need to generate in order to simulate 100 days in which 35 seats were booked?

Chapter 1 Summary

What have we learned?

The purpose of this chapter is to gain a better understanding of the nature of random behavior. Random behavior means that we do not know what will happen in the future. Often we know one of a number of possibilities will happen. However, we do not know which of them will occur. One way to study situations that include multiple random events is to simulate them. We started by simulating two equally likely outcomes by flipping coins. When we have outcomes that are not equally likely we use a random number generator on a calculator or computer. The purpose of the simulation is to try to discover what would happen if a situation occurred a large number of times. This can give us an idea of how likely a particular outcome is. We can use this information to make decisions.

We have also learned how to combine the probabilities of two or more events if we know the probability of each of them occurring. For independent events, we can multiply the probabilities together to find the probability that they both occur.

Finally, we learned how to use an Excel spreadsheet for larger scale simulations. Excel can be used for generating random numbers and can also do calculations that can be used to evaluate the results. When we flip coins or use a calculator to generate the results, we have to write the results down. Excel can keep and easily display all of the simulated results. It is important to remember to use the paste special "values" command to save the numbers generated. This is to prevent random numbers that were generated from changing the next time we enter a command.

Our intuition when it comes to random behavior is often wrong. People have a hard time getting rid of the belief that the future will correct for a surprising run of random behavior in the past. There is no "Law of Averages" that states that the outcomes of a random event will "even out" within a small sample. However, the average of the results of a large number of trials will get closer to the expected value as an event is repeated. This is not because future results somehow make up for the past. It is simply that a small number of results have a small impact on the average of a large number of results.

Terms

Chance event — Events whose outcomes are random.

Collectively exhaustive events — A set of events is collectively exhaustive if at least one of them must occur (i.e., a collectively exhaustive set of events includes all possible outcomes).

Collectively exhaustive sets — If sets A and B are collectively exhaustive subsets of the universal set U, then $A \cup B = U$.

Complement of a set — The complement of set A, A', contains all of the elements that are not included in set A. The shaded region of the Venn diagram below shows the complement of A. The complement of set A is often written as A^c or A' or $\overline{A}$.

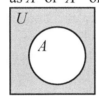

Complementary events and probability — Two events A and B are complementary if they are mutually exclusive and collectively exhaustive.

$$P(A) + P(B) = 1$$
$$P(B) = 1 - P(A)$$
$$P(A) = 1 - P(B)$$

Compound events — A compound event consists of two or more simple events.

Element of a set — The objects that make up a set are referred to as elements of the set. If x is an element of set A, then $x \in A$, and if y is not an element of set A, $y \notin A$, as shown in the Venn diagram below.

Empty (or null) set — The empty set is the set containing no elements, denoted $\emptyset$ or { }.

Event — Any subset of the sample space is called an event.

Experiment — An experiment is an activity under consideration whose outcome is left to chance.

Independent events and probability	Two events A and B are independent if the outcome of one event does not affect the outcome of the other. If A and B are independent, then $$P(A \text{ and } B) = P(A \cap B) = P(A) \cdot P(B)$$
Intersection of sets	The intersection of sets A and B contains all the elements that belong to set A and also to set B. The shaded region of the Venn diagram below shows the intersection of sets A and B. Notice that A and B have this region in common. 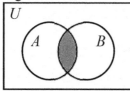
Mutually exclusive events and probability	Two events are mutually exclusive if the occurrence of one event precludes the possibility of the other event occurring. If events A and B are mutually exclusive, then the probability of event A or event B occurring is the sum of the probabilities of events A and B. $$P(A \text{ or } B) = P(A \cup B) = P(A) + P(B)$$
Mutually exclusive sets	If sets A and B are mutually exclusive subsets of the universal set U, then they have no elements in common. Two sets that are mutually exclusive can also be described as disjointed sets. 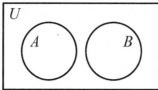
Outcome	Each possible observation or occurrence in an experiment is called an outcome.
Probability	The probability of an event is a measure of the likelihood of that particular event occurring. The probability of an event is a fraction or decimal between zero and one. Events that are unlikely have probabilities closer to zero. Events that are likely to occur have probabilities closer to one. If all possible outcomes are equally likely, the probability of an event occurring equals the ratio of the number of outcomes that result in a particular event to the total number of possible outcomes in the sample space. The probability P of event E occurring can be written using the following notation: $$P(E) = \frac{\text{number of favorable outcomes}}{\text{total number of possible outcomes}}.$$

Random process	A random process models changes in a system over time, where the changes lack a definite plan, purpose, or pattern.
Random variable	A random variable is an expression whose value is subject to chance variations lacking any definite plan, purpose, or pattern. Like any other variable in mathematics, a random variable can take on different values.
Randomness	Randomness implies that the outcome of an event does not follow any prescribed pattern and is thus unpredictable. Another way of saying this is that the outcome of the event occurs strictly by chance.
Observed relative frequency	The ratio of the number of times an event is observed to occur to the total number of observations. The observed relative frequency of an event can be used as an estimate of the probability of the event. The observed relative frequency of an event is also referred to as the empirical probability of the event.
Outcome	The result of an experiment or simulation involving uncertainty.
Sample space	The sample space is the set consisting of all the possible outcomes of an experiment. This set is often denoted S.
Sequence	A list of outcomes when an event is repeated.
Set	The term set is undefined, but may be thought of as a collection of objects. In set-builder notation, the elements of a set are enclosed in braces; e.g., $S=\{0, 1, 2\}$.
Simulation	A simulation is a process used to model random events.
String	A sequence of consecutive results with the same outcome
Subset	A subset of a set contains elements, all of which are contained in another set. If B is a subset of A, then every element of B is also an element of A, denoted by $B \subseteq A$. The Venn diagram below shows the subset relationship where B is completely contained within A; i.e., B is a subset of A.

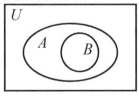

Union of sets   The union of set *A* and set *B* contains all of the elements that belong to set *A* or set *B* or *both* set *A* and set *B*. In this sense, we are using the word *or* inclusively, which means either or both. The shaded region of the Venn diagram below shows the union of sets *A* and *B*.

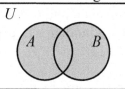

Universal set   The set containing all possible elements is denoted as *U*.

Variability   Variability is a characteristic of data that refers to the fact that each data value is not necessarily the same.

Chapter 1 (Basic Probability and Randomness) Objectives

You should be able to:

- Identify the possible outcomes for an event.

- Determine the probability of each of the possible outcomes for an event.

- Determine whether or not two events are complementary.

- Use the concept of complementarity to calculate the probability of an event.

- Use the multiplication principle to determine the probability of two independent events both occurring.

- Use "randInt" on the TI calculator and "randbetween" in Excel to generate a random integer between two values.

- Use "countif" function in Excel to find out how often a particular value appears in a list.

- Evaluate the results of a simulation to find the likelihood of a particular outcome.

Chapter 1 Study Guide

1. What is the difference between independent events and mutually exclusive events? Give an example of each.

2. What does it mean if two events are complementary? Give an example.

3. Give an example of two events that are mutually exclusive but not complementary.

4. Assume two events are independent and both have a 90% chance of occurring.

    a. What is the probability both will occur?

    b. What is the probability one will occur and one will not?

    c. What is the probability neither will occur?

    d. If a third independent event also has a 90% chance of occurring, what is the probability of all three occurring?

5. What does the notation $P(A)$ mean? What about the notation $P(A^c)$?

6. If $P(A) = 0.75$ what is $P(A^c)$?

7. Are the results of a simulation always reliable? Explain.

CHAPTER 2:

Conditional Probability

Section 2.0 Conditional Probability

Conditional probability is a critical concept within probability theory. In essence it means that the probability of an event occurring can be influenced by another piece of information. We have already discussed the fact that many people have poor intuitions when probability, uncertainty, and randomness are involved. This is especially true with regard to the concepts of conditional probability.

The basic notation for conditional probability is shown below.

$$P(B \mid A)$$

This is read as the probability of B occurring given that A is known or has occurred. Knowing that event A has occurred sometimes allows you to calculate a different probability of event B occurring, simply because you now have more information.

2.0.1 Mortality Tables

One classic example of conditional probability involves the probability distribution of future life for people of different ages. The Social Security Administration (SSA) estimate the average lifetime of a male born in 2011 was a little more than 76 years. Poor intuition leads many people to imagine a male who has reached 80 years of age is living on borrowed time. In fact, the SSA estimates the probability of an 80 year old dying within his next year is 0.06. On average, he will live more than eight additional years. Table 2.0 presents actuarial data maintained by the Social Security Administration of the US. The table starts with a cohort of a 100,000 of each gender. It presents a forecast of the number of people in each cohort who will live to a specific age. This table stops at the age of 20. https://www.ssa.gov/oact/STATS/table4c6.html

Exact age	Number of Lives			Exact age	Number of Lives		
	Males	Females	Total		Males	Females	Total
0	100,000	100,000	200,000				
1	99,343	99,449	198,792	11	99,155	99,296	198,451
2	99,299	99,411	198,710	12	99,146	99,288	198,434
3	99,270	99,389	198,659	13	99,132	99,277	198,409
4	99,248	99,373	198,621	14	99,112	99,265	198,377
5	99,230	99,358	198,588	15	99,080	99,248	198,328
6	99,215	99,346	198,561	16	99,037	99,228	198,265
7	99,200	99,334	198,534	17	98,983	99,204	198,187
8	99,187	99,324	198,511	18	98,917	99,175	198,092
9	99,175	99,314	198,489	19	98,837	99,144	197,981
10	99,164	99,305	198,469	20	98,744	99,109	197,853

Table 2.0.1: 2011 Social Security Administration actuarial data for the US

Of a total of 100,000 male newborns, the table predicts that 98,744 males will still be alive at age 20. Alternatively, it forecasts that 1,256 of the 100,000 males will have died before the age of 20. In contrast, the number of survivors in a female cohort is forecasted to be 99,109. The corresponding number of deaths would be 901. Now consider a combined population of 200,000 newborns that are equally divided between males and females. The total number of forecasted deaths is 2,147.

This data can be converted into probability statements.

Let: D = the age of individual at death
M = individual is male
F = individual is female

The probability of a randomly selected individual in the total 200,000 will die by age 20 is
$$P(D<20) = (2147/200,000) = 0.0107 \quad \text{or} \quad \text{about 1 chance in 93}$$

However, this probability is different for males and females. The probability of death by age 20 for a female infant is 0.0090.

$$P(D<20|F) = 0.0090 \quad \text{or} \quad \text{about 1 in 111}$$

The corresponding probability of death for a male infant is 0.0122. There is a 36% higher probability than for a female.

$$P(D<20|M) = 0.0124 \quad \text{or} \quad \text{about 1 in 80}$$

Knowing the gender of an infant clearly affects the probability of death by age 20. Thus, death by age 20 is not independent of an individual's gender.

The concept of conditional probability can be applied another way to the data in Table 2.1. The probability of dying by age 20 can be conditioned on the individual's age at the time.

Let A = the age of an individual.

$$P(D<20|A)$$

In the total cohort, there are 198,328 individuals of age 15. Of that total 197,853 will survive to age 20. Equivalently, the number of deaths in this group is forecasted to be 475. Thus,

$$P(D<20|A=15) = \frac{475}{198,328} \approx 0.0024 \quad \text{or} \quad \text{about 1 in 418}$$

This probability is much smaller than the corresponding probability for a newborn which is 0.0124. Thus a youngster's current age and death by age 20 are not independent events.

All of life's risks can be adjusted based on information we know about the individual. This information could be the gender, race, lifestyle, location, age, etc. The likelihood of dying from lung cancer is very different for a smoker and a non-smoker. Similarly, the likelihood of developing diabetes is not the same for a person of average weight as compared to someone who is obese. The likelihood of being a victim of a violent crime is much lower for someone living in an affluent suburb than in a much less affluent inner city neighborhood.

1. Provide your own example of a risk people of your age might face that will vary based on some other given information.

Chapter 2 — Conditional Probability

Section 2.1 Committee Diversity

The concept of ***independent events*** can be defined using conditional probability. Two events are said to be independent of one another if information about one event does not affect the probability of the other event. This statement is represented as

$$P(B|A) = P(B) \leftrightarrow A \text{ and } B \text{ are independent events and}$$
$$P(A|B) = P(A) \leftrightarrow A \text{ and } B \text{ are independent events.}$$

The double-headed arrows ($\leftrightarrow$) in the statements above are read "if and only if" and show the equivalence of the statements on either side of the arrows.

Flipping a coin is good example of independent events.

Let H_1 first coin flip is heads
   H_2 second coin flip is heads

The likelihood of a coin flip coming up heads is 0.5 on the first flip. The likelihood of the coin flip coming up heads is 0.5 on the second flip. This probability is not affected by what happened on the first flip.

$$P(H_2 | H_1) = P(H_2) = 0.5$$

Therefore, H_2 and H_1 are independent events.

In contrast, consider a group of four boys and three girls who have all volunteered to serve on a trip planning committee for Ms. Doubtful's class. Ms. Doubtful wondered what might happen if she selected the two without paying any attention to their gender. If two students are selected at random by drawing names out of a hat, what is the likelihood that both are boys? Conditional probability is helpful in answering this question by analyzing the sequence of random events.

The probability that the first person selected is a boy is just the ratio of boys to the total group.

$$P(B_1) = \frac{4}{7}$$

If the first person selected is a boy, there are still three boys among the group of six remaining students.

$$P(B_2 | B_1) = \frac{3}{6}$$

If, however, a girl was picked first, there are four boys remaining among the group of six students.

$$P(B_2 | G_1) = \frac{4}{6}$$

Clearly, the likelihood that the second selection is a boy changes depending on what happened in the first selection. Now to return to original question, what is the likelihood of selecting two boys? This can be expressed as $P(B_1 \cap B_2)$. If these were independent events, Ms. Doubtful could use the multiplication rule and multiply two probabilities. However, they are not independent. The general form for determining the probability of the intersection of two events is

$$P(A \cap B) = P(A) \cdot P(B | A)$$

If A and B are independent then $P(B|A) = P(B)$. As a result, the more general conditional probability formula reduces to the multiplication formula for independent events.

$$P(A \cap B) = P(A) \cdot P(B | A)$$
$$= P(A) \cdot P(B)$$

In this example,

$$P(B_1 \cap B_2) = P(B_1) \cdot P(B_2 | B_1)$$
$$= \frac{4}{7} \cdot \frac{3}{6}$$
$$= \frac{12}{42}$$
$$\approx 0.29$$

This formula suggests that the intersection of two events can be analyzed by considering them as a sequence. First determine the likelihood of the first event. Then determine the likelihood of the second event *conditioned on what happened first*.

Ms. Doubtful applied the same logic to determine the likelihood that both will be girls?

$$P(G_1 \cap G_2) = P(G_1) \cdot P(G_2 | G_1)$$
$$= \frac{3}{7} \cdot \frac{2}{6}$$
$$= \frac{6}{42}$$
$$\approx 0.14$$

It is twice as likely that the committee will consist of two boys, 12/42, as compared to the probability that the committee will consist of two girls, 6/42. Ms. Doubtful thought the planning committee would better represent the class's interest if there were one boy and one girl on the committee. However, she was not prepared to force this to happen. She wanted to give each person who volunteered the same opportunity to be selected.

To calculate the probability of a balanced committee, she reasoned there were two ways this could occur. It could happen that the first student selected was a boy and the second was a girl and vice versa. These two sequences are mutually exclusive; therefore, their probabilities can be added.

$$\begin{aligned} P(\text{one boy and one girl}) &= P(B_1 \cap G_2) + P(G_1 \cap B_2) \\ &= P(B_1) \cdot P(G_2 | B_1) + P(G_1) \cdot P(B_2 | G_1) \\ &= \frac{4}{7} \cdot \frac{3}{6} + \frac{3}{7} \cdot \frac{4}{6} \\ &= \frac{12}{42} + \frac{12}{42} \\ &= \frac{24}{42} \\ &\approx 0.57 \end{aligned}$$

The equation above represents the concept of a **partition**. This involves decomposing an outcome into all of its mutually exclusive ways of happening. In this example B_1 and G_1 are a partition of the outcome of the first student selected. Ms. Doubtful calculated the probability of selecting one girl and one boy by partitioning on the set of all possible outcomes of the first pick.

Another method for determining this probability involves using the concept of complementary events.

2. Use the probabilities calculated for an all boys' committee or an all girls' committee to determine the likelihood of a mixed committee.

It was more likely that the committee would have one boy and one girl than two of the same gender. However, Ms. Doubtful would like to have better odds of there being at least one boy and one girl on the committee. Danielle Wiseman, the class math whiz, suggested that she expand the committee to three people.

Ms. Doubtful used the same logic as before to calculate the chances of all three members being of the same gender. She imagined a sequence of selections. The first person selected is a boy and then the second is a boy and then the third is a boy. She applied the same logic of conditional probability with one modification. When it came to the third boy, she had to condition on what had happened on the first two selections. She needed to determine $P[B_3|(B_1 \cap B_2)]$. The resultant formula was

$$P(B_1 \cap B_2 \cap B_3) = P(B_1) \cdot P(B_2 \mid B_1) \cdot P[B_3 \mid (B_1 \cap B_2)]$$
$$= \frac{4}{7} \cdot \frac{3}{6} \cdot \frac{2}{5}$$
$$= \frac{24}{210}$$
$$\approx 0.11$$

Ms. Doubtful asked Danielle to complete the analysis.

3. What is the probability that the committee would consist of three girls?

4. What is the probability that the committee would be mixed and include at least one girl and at least one boy? (Use the concept of the complement to calculate this probability.)

In the calculations above, we presented the analysis as a sequence of two events. First we pulled one name out of the hat, and then we picked a second name. Hopefully, it is clear that the probabilities calculated above would also apply if Ms. Doubtful put her hand in the hat and selected two names at once. The manner in which the two names are picked should not affect the probabilities. This illustrates an important point. The logic of conditional probability can be used to determine a joint probability even if the events do not occur in sequence.

In this example, each of the four conditional probabilities could be calculated using simple logic. In other contexts, the conditional probabilities are obtained through data analysis. This is demonstrated in the next example.

Section 2.2 Pairs of Free Throws in Basketball

Blu Bauerbach is the coach of the Wellington Dukes girls' varsity basketball team. Coach Bauerbach frequently stops practice and has her team line up for free throw practice. Each player takes two shots, and the coach keeps careful records of makes and misses. She is interested in knowing whether making the first shot causes the girls to be more likely to make the second shot. Blu is also interested in what happens if the player misses her first shot. In essence, she is interested in determining conditional probabilities. How likely is it that a player who just made a shot will make her next one?

Susan Olafsen is the starting center for the Dukes.

Let X_1 = the event that Susan made the first free throw,

X_2 = the event that Susan made the second free throw,

O_1 = the event that Susan missed the first free throw, and

O_2 = the event that Susan missed the second free throw.

The conditional probabilities of interest to Coach are represented using the following notation.

$$P(X_2 \mid X_1)$$

This notation should be interpreted as "the probability of event X_2 occurring, given that event X_1 has occurred." More succinctly, it can be read "the probability of X_2, given X_1." In this context, it is the probability of making the second free throw given she made the first free throw.

Table 2.1.1 shows the data Coach Bauerbach recorded for Susan. In the table, the letter X represents a made shot and O represents its complement, a missed shot. The data consists of 30 pairs of free throws.

	Shot			Shot			Shot	
Pair	1	2	Pair	1	2	Pair	1	2
1	X	O	11	X	O	21	X	X
2	X	X	12	O	X	22	X	X
3	X	O	13	X	X	23	X	O
4	X	X	14	X	X	24	X	O
5	X	O	15	X	X	25	O	X
6	X	O	16	X	O	26	O	X
7	X	X	17	O	X	27	X	X
8	O	O	18	X	X	28	X	X
9	O	X	19	X	O	29	X	X
10	X	O	20	X	X	30	X	O

Table 2.2.1: Data from Susan Olafsen's free throw shooting (X = make, O = miss)

Coach Bauerbach wants to know how likely it is Susan will make the second free throw based on whether or not the first free throw went in. All the data in Table 2.2.1 needs to be summarized to make these probabilities more apparent.

The first probability of interest was Susan's likelihood of making the first free throw. In the data set, it was simpler to count the number of misses. She missed her first free throw only six times out of 30.

$$P(O_1) = \frac{6}{30} = 0.2$$

1. What is the probability of making her first free throw?

Conditional probability shrinks the sample space. Instead of considering all possible outcomes, we only need to concern ourselves with outcomes involving X_1. Susan made her first free throw 24 times. She made both free throws 13 times. Thus, given she made her first free throw, the proportion of times she made her second free throw was 13 out of 24.

$$P(X_2 \mid X_1) = \frac{13}{24} \approx 0.54$$

The event X_1 corresponds to the shaded circle in the Venn diagram. The conditional probability corresponds to the ratio of the overlapping portion to the shaded circle.

2. Add numbers to each of the four regions in the Venn diagram below to reflect the number of makes and misses on both the first and second shot for the 30 pairs of free throws.

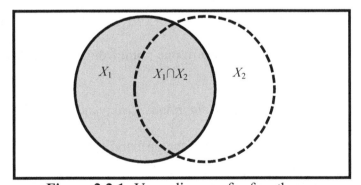

Figure 2.2.1: Venn diagram for free throws

We can also use the general formula for calculating the probability of making both free throws. Not surprisingly, the answer is 13/30.

$$P(X_1 \cap X_2) = P(X_1) \cdot P(X_2 \mid X_1)$$
$$= \frac{24}{30} \cdot \frac{13}{24}$$
$$= \frac{13}{30}$$
$$\approx 0.43$$

3. Use the data to determine the probability Susan will miss both free throws.

This general formula for the intersection can be manipulated to calculate conditional probabilities.

$$P(X_2 \mid X_1) = \frac{P(X_2 \cap X_1)}{P(X_1)}$$
$$= \frac{\frac{13}{30}}{\frac{24}{30}}$$
$$= \frac{13}{24}$$
$$\approx 0.54$$

4. Use this formula to determine $P(O_2 \mid O_1)$.

5. How many times did Susan miss the first free throw?

6. Within this subset of events, how many times did Susan make the second free throw?

7. What is $P(X_2 \mid O_1)$?

There are two different ways that Susan can earn one point from two free throws; $\{X_1 \cap O_2\}$ or $\{O_1 \cap X_2\}$.

8. Use the data in the table to determine the relative frequency of each of these two events.

9. Use conditional probability to determine the probability of each of these two events? Explain why these two probabilities are not equal.

2.2.1 Random Variable and Expected Value

In this analysis we considered each of the four possible outcomes:

$\{X_1 \cap X_2,\ X_1 \cap O_2,\ O_1 \cap X_2,\ O_1 \cap O_2\}$

Often, we are just interested in the number of free throws made and not the actual pattern. When we convert the detailed sample space into a numerical value, we are creating a ***random variable***. A random variable is a variable whose value is subject to variations due to chance.

Let $Y = 0$ if $O_1 \cap O_2$ occurs,

$Y = 1$ if $X_1 \cap O_2$ or $O_1 \cap X_2$ occurs, and

$Y = 2$ if $X_1 \cap X_2$ occurs.

The probability of each value corresponds to the probabilities of the outcomes that make up that value.

$$P(Y=0) = P(O_1 \cap O_2)$$
$$= \frac{1}{30}$$

$$P(Y=1) = P(X_1 \cap O_2) + P(O_1 \cap X_2)$$
$$= \frac{11}{30} + \frac{5}{30}$$
$$= \frac{16}{30}$$

$$P(Y=2) = P(X_1 \cap X_2)$$
$$= \frac{13}{30}$$

The ***expected value*** of a random variable is defined as the *weighted sum* of all of the possible outcomes. Each outcome is weighted by its probability.

$$E(Y) = \sum [Y \cdot P(Y)]$$
$$= 0 \cdot P(Y=0) + 1 \cdot P(Y=1) + 2 \cdot P(Y=2)$$
$$= 0\left(\frac{1}{30}\right) + 1\left(\frac{16}{30}\right) + 2\left(\frac{13}{30}\right)$$
$$= \frac{42}{30}$$
$$\approx 1.4 \text{ points}$$

It is unfortunate that the use of the word "expected" in *expected value* is not consistent with the way we use "expected" in normal speech. When we say we expect to graduate high school or pass an exam, we mean that this outcome is very likely to happen. We *expect* it to happen.

However, the expected value is NOT expected to happen. The expected value of the number free throws Susan will make is 1.4. There is no way this can actually happen. The expected value corresponds to the long-range average value that will be observed. For example, if Susan were to shoot 200 pairs of free throws, the average number of makes per pair would be close to 1.4. (How close to 1.4 can be calculated using advanced concepts of statistics.)

To understand this concept of long range average, imagine you were tossing a single fair die. Each one of the six sides is equally likely to occur. Each probability is 1/6. The expected value

$$E(Y) = 1\left(\frac{1}{6}\right) + 2\left(\frac{1}{6}\right) + 3\left(\frac{1}{6}\right) + 4\left(\frac{1}{6}\right) + 5\left(\frac{1}{6}\right) + 6\left(\frac{1}{6}\right)$$
$$= \frac{21}{6}$$
$$\approx 3.5$$

Needless to say, it is impossible to roll 3.5 on a single die. However, imagine you were to roll the dic thousands of times and record each observation. The average of all of these outcomes would be close to 3.5.

In our example, we calculated all of the probabilities using the relative frequency observed in the data set in Table 2.1.2. Thus, the expected value and average will be identical. (See Table 2.2.2.) In total she would have scored 42 points with these 30 pairs of free throws for an average of 1.4 points per pair.

Pair	Shot 1	Shot 2	Points	Pair	Shot 1	Shot 2	Points	Pair	Shot 1	Shot 2	Points
1	X	O	1	11	X	O	1	21	X	X	2
2	X	X	2	12	O	X	1	22	X	X	2
3	X	O	1	13	X	X	2	23	X	O	1
4	X	X	2	14	X	X	2	24	X	O	1
5	X	O	1	15	X	X	2	25	O	X	1
6	X	O	1	16	X	O	1	26	O	X	1
7	X	X	2	17	O	X	1	27	X	X	2
8	O	O	0	18	X	X	2	28	X	X	2
9	O	X	1	19	X	O	1	29	X	X	2
10	X	O	1	20	X	X	2	30	X	O	1

Table 2.2.2: Points scored from Susan Olafsen's free throw shooting

2.2.2 Susan Olafsen at the Line to Shoot One-and-One

Imagine the Dukes are down by one point with less than one second to play. Just before the buzzer ending the game, Susan Olafsen was fouled before shooting. She will be shooting one-and-one free throws. She gets a second free throw only if she makes the first. Use the original probabilities found earlier to answer the following questions.

1. What is the probability the Dukes will win the game?

2. What is the probability the Dukes will lose the game?

3. How likely is it that the game will go into overtime (i.e., the game will be tied when regulation time expires)?

4. In a one-and-one situation, what is the expected value of the number of points scored? Explain why this not the same as when Susan is taking two free throws as a result of being fouled in the act of shooting.

The pressure of a situation can affect someone's performance. Imagine that Susan's first shot success rate declines to 0.65 when the game is on the line. (Assume her second shot is not affected by the pressure.)

5. What is the probability the Dukes will win the game?

6. What is the probability the Dukes will lose the game?

2.2.3 Improved Performance

Coach Blu Bauerbach is concerned that after making her first free throw, Susan makes her second shot only 54% of the time. This is significantly lower than the 80% probability on the first free throw. She is concerned that Susan may not be focusing enough after succeeding the first time. Blu has Susan perform the following practice. Take five shots from the field. Then take two free throws. Repeat this cycle of regular shots 10 times. Her goal is to have Susan's performance on the second shot to be the same and independent of her performance on the first shot. Her goal is for Susan to make 80% of her second shots.[1]

7. If Susan achieved this goal, what would be the probability of making both free throws, $P(X_2 \cap X_1)$?

8. What is the expected value of the number of points scored? (Hint: It is not 1.4)

Imagine the one-and-one situation described before. (Ignore the comments about performance under pressure.)

9. What is the probability the Dukes will win the game?

10. What is the probability the Dukes will lose the game?

11. Explain any differences you find compared to the answers to questions 1 and 2.

---

[1] Independence of the first and second shots does not necessarily mean equal probabilities. If she made 70% of her second shots irrespective of her first shot, the two events would be independent. Similarly, if the first shot helps her focus, she might make 90% of her second shots independent of how well she did on the first.

Section 2.3 Home Court Advantage in the NBA

Conditional probability can be easily understood in the context of a simple table. Table 2.2.1 lists the win and loss records for four NBA teams along with two fictitious teams. These data are from the middle of April 2014 when each team had played 79 or 80 games. The season is 82 games long. The winning percentage of the Indiana Pacers was 67.5%. This percentage is an estimate of the probability that it will win a game it plays. However, if you know the game is to be played on its home court, the probability of winning increases to 0.85. If it is an away game, the probability is only 0.50. The Atlanta Hawks had an overall losing record. Yet at home, they won 59% of their games.

Team	April 11-13, 2014								
	Home			Away			Total		
Indiana Pacers	34	6	85.0%	20	20	50.0%	54	26	67.5%
Atlanta Hawks	23	16	59.0%	13	27	32.5%	36	43	45.6%
GS Warriors	26	14	65.0%	23	16	59.0%	49	30	62.0%
OKC Thunder	33	7	82.5%	25	15	62.5%	58	22	72.5%
Maine Crabs	35	6	85.4%	19	20	48.7%	54	26	67.5%
San Diego Seals	33	6	84.6%	21	20	51.2%	54	26	67.5%

Table 2.3.1: NBA teams' wins and losses for home and away games

10. Add numbers to each of the four regions in the Venn diagram below to reflect the number of home and away games won and lost by the Indiana Pacers.

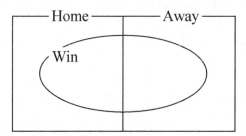

Figure 2.3.1: Venn diagram for Indiana's wins and losses

The concept of conditional probability can be represented as follows.

Let I_w = Indiana wins a game,

I^c = Indiana loses a game,
H = the game is played at home, and
A = the game is played away from home.

The unconditional probability of Indiana winning a randomly selected game is

$$P(I_w) = 0.675.$$

However, conditioned on the location of the game, the probability of winning changes.

$P(I_w | H) = 0.85$ and
$P(I_w | A) = 0.50$.

11. What are the unconditional and conditional probabilities of winning for the Oklahoma City Thunder?

12. Which team has the largest difference in conditional probabilities for winning home and away games?

13. Which team has the smallest difference in conditional probabilities for winning home and away games?

There are two games left in the season, one away and one at home. Indiana needs to win both games to insure that it earns home court advantage in the first round of the playoffs.

14. What is the probability Indiana wins both games?

We have created two additional hypothetical teams with similar overall records as Indiana. The Maine Crabs have two away games left. The San Diego Seals have two home games left.

Calculating the probability that San Diego wins its last two home games involves an additional complication. San Diego has won 33 out of 39 home games, a winning percentage 0.846. After San Diego wins the first game, its winning percentage is slightly improved. Its home winning percentage increases to 0.85 (34/40).

Let S_i = San Diego wins Game i
H_i = Game i is played at home

$P(S_1 | H_1) = 33/39 \approx 0.846$

$S_1 \cap H_1 \cap H_2$ represents the intersection of three events. San Diego won the first game (S_1), the first game was at home (H_1), and the second game is also to be played at home (H_2).

$P[S_2 | (S_1 \cap H_1 \cap H_2)] = 34/40 = 0.85$

The above notation seems complicated. It represents the probability that San Diego wins the second game, given it won the first game and both games were played at home. The joint probability of winning both games involves multiplying the above two formulas.

$$P[(S_1 \cap S_2)|(H_1 \cap H_2)] = P(S_1|H_1) \cdot P[S_2|(S_1 \cap H_1 \cap H_2)]$$
$$= \frac{33}{39} \cdot \frac{34}{40}$$
$$= (0.846)(0.85) \approx 0.719$$

15. What is the probability that Maine wins its last two games?

You should have found that there are large differences in the probability of winning the final two games of the season for different schedules. It is, therefore, noteworthy that the NBA tries to schedule the last two games of the regular season as one home and one away game for each and every team.

Section 2.4 Defective Ignition Parts at a Manufacturing Plant

The Hot Rider Motorcycle Inc. purchases its electronic ignitions from two different suppliers, Cheapo and Tres. Cheapo has a defect rate of four ignitions per 100 shipped. It charges $40 per ignition. Tres has better quality control and its defect rate is one per 100. Tres charges $50 for its higher quality ignitions. These defect rates are averages per 100 ignitions. In any batch of 100 there may be more or less than this average. To save money, Hot Rider currently purchases 80% of its ignitions from Cheapo.

The nature of the defect is such that is extremely difficult to identify the problem before it is installed in a motorcycle. Vanna Gogh is director of warranty claims and repairs. She has found the defective ignition typically fails between 500 and 1,000 miles of motorcycle riding. Vanna is interested in determining the probability that a randomly installed ignition is defective.

This probability can be computed using the concept of a **partition**. A partition divides a set into mutually exclusive and collectively exhaustive sets. In this instance all ignitions come from either Cheapo or Tres. The sets C and T do not overlap and are therefore, **mutually exclusive**. In addition all parts come from one of these two suppliers. They are therefore **collectively exhaustive**. The concept of a partition can be used to calculate the probability of an event by combining it with the other elements of the partition.

Imagine that Hot Rider has 800 Cheapo ignitions and 200 Tres ignitions. Assume that exactly 4% of the Cheapos are defective and exactly 1% of the Tres are defective.

1. Add numbers and label each of the four regions in the Venn diagram below to reflect the number of defective and non-defective ignitions from each company in this group of 1,000 ignitions. Let the shaded circle represent all of the defectives.

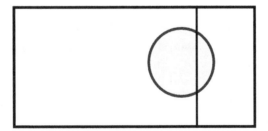

Figure 2.4.1: Venn diagram for defective ignitions

Let D = part is defective
         C = part is from Cheapo
         T = part is from Tres

$$P(D|C) = 0.04$$
$$P(D|T) = 0.01$$
$$P(C) = 0.8$$
$$P(T) = 0.2$$

Each defective part will come from either Cheapo or Tres.

$$P(D) = P(D \cap C) + P(D \cap T)$$
$$= P(C \cap D) + P(T \cap D)$$

The intersection of D and C can be calculated using conditional probability.

$$P(C \cap D) = P(C) \cdot P(D|C)$$

The same is true for the intersection of D and T.

$$P(T \cap D) = P(T) \cdot P(D|T)$$

Thus, the overall probability that an ignition will be defective is

$$P(D) = P(C \cap D) + P(T \cap D)$$
$$= P(C) \cdot P(D|C) + P(T) \cdot P(D|T)$$
$$= (0.8)(0.04) + (0.2)(0.01)$$
$$= 0.034$$

This corresponds to the 34 defective parts out of 1,000 that should have been placed in the circle of the Venn diagram.

2.4.1 Reduce the Defect Rate

Vanna Gogh is concerned about the defect rate. Every time the ignition fails, a truck must be sent to pick up the motorcycle. It then has to be repaired and returned to the owner. The pickup and repair cost can be more than $150. In addition, customers are upset about what happened and may bad-mouth the product. She would like to reduce the defect rate to less than 2 per 100 without adding too much cost. In other words, Vanna would like the probability of an ignition failure to be less than 0.02.

She is considering changing the purchasing strategy. For example, she wondered what would happen if purchases were split equally between Cheapo and Tres. The current cost of buying 1,000 ignitions is

$$\$40 \cdot 800 + \$50 \cdot 200 = \$32,000 + \$10,000 = \$42,000.$$

However, the overall probability of a defect with her new purchasing plan would still be more than 0.02.

$$P(D) = P(C \cap D) + P(T \cap D)$$
$$= P(C) \cdot P(D|C) + P(T) \cdot P(D|T)$$
$$= (0.5)(0.04) + (0.5)(0.01)$$
$$= 0.025$$

Vanna knows that if she buys only Tres ignitions, she will meet her goal. However, the total cost of that plan would be $50,000.

2. Identify a purchasing policy that achieves the standard but costs thousands of dollars less than buying all parts from Tres.

3. Determine the probability of a defective ignition and the total purchase cost for 1,000 ignitions.

Another option is to hire a company that can rigorously test the ignitions for failures. The test will cost $7 per ignition. In addition the test is not perfect. The probability of identifying a defective ignition is only 0.7. This means that 30% of the defective ignitions will still get through.

Let D = part defective
    I = installed in motorcycle

The probability that a defective Cheapo ignition will be installed is calculated below.

$$P(D \cap I) = P(D) \cdot P(I|D)$$
$$= (0.04)(0.3)$$
$$= 0.012$$

4. Determine the probability that a defective Tres ignition will be installed.

Assume that Hot Rider Motorcycle hires the testing company and continues to purchase 80% of it ignitions from Cheapo and 20% from Tres.

5. Determine the probability that a randomly selected motorcycle will have a defective ignition.

6. What is the total cost of using this policy to purchase and test 1,000 ignitions?

Patsy Jack, Vanna's assistant, had an interesting idea. Perhaps it was not necessary to test Tres ignitions, because they are 99% reliable.

7. Would this policy achieve the goal of less than a 0.02 defect rate? What would this policy cost?

Section 2.5 College Graduation Rates and ACT Scores

Donald Williams and Serena Trump were exploring their options for college after their graduation from Towers High School. Donald had been a B+ student in high school; Serena had been a B/B+ student. Neither student performed well on standardized tests. Donald earned a combined score of 20 on the ACT. Serena performed a little better; her score was 23. They enjoyed their years at Towers, but they recognized that their school was not highly rated academically. They each knew many older alumni who had gone off to college with high hopes. More than 40% of them, however, were unable to complete a four-year college in fewer than six years. Many had dropped out after the first year or two.

Donald and Serena had both been accepted to Warren University and Taft University. One factor they wanted to consider was the graduation rate of these two schools. A quick scan of the Internet indicated that Warren University had the higher six-year graduation rate. Its graduation rate was 61% as compared to 58% for Taft. Donald contacted Warren University to find out more about the school's efforts to help students graduate within six years. Serena did the same with Taft University. After comparing notes, they were puzzled by what they found. In talking with staff, it was clear that Taft University provided more resources to help students succeed. If that were true, why was Warren University's graduation rate higher than Taft's? They went to see Ms. Venus Stats, their math teacher and guidance counselor.

Ms. Stats had access to more detailed graduation rate data for each school. She provided Donald and Serena with the data in Table 2.4.1. The data showed the graduation rate for different ACT score ranges. After looking at the data, they were even more perplexed. Taft seemed to be better in many categories and almost as good in the other categories; yet, in total it was worse.

ACT Score Range		Warren University		Taft University	
		Graduation percent	Students	Graduation percent	Students
R_1	16-18	45%	5%	50%	25%
R_2	19-21	52%	15%	58%	30%
R_3	22-24	59%	30%	62%	35%
R_4	25-27	65%	35%	64%	5%
R_5	28-30	71%	15%	70%	5%
	Aggregate	61%		58%	

Table 2.5.1: ACT, graduation, and enrollment data for Warren and Taft

Ms. Stats began by suggesting that visualizing the data often can help the numbers make sense. She produced the bar chart in Figure 2.4.1 and pie charts in Figure 2.4.2. Pie charts are useful here, because the ACT score ranges partition the set of students at Warren and Taft Universities. The pie charts, therefore, double as a Venn diagram of the subsets while also showing the relative sizes of the subsets.

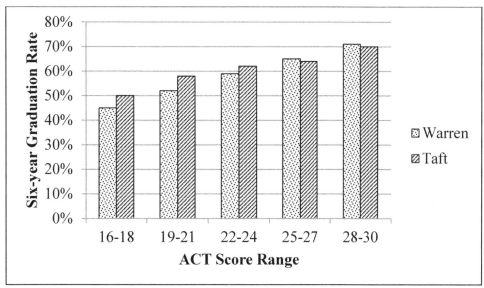

Figure 2.5.1: Comparison of six-year graduation rates versus students' ACT scores

The chart in Figure 2.4.1 shows the graduation rates of students in the various ranges of ACT scores. There does not appear to be a great difference between the two universities based on these data. Taft performs better in three of the ranges by several percentage points. Warren performs better in two ranges by just one percentage point. Then Ms. Stats showed them the charts comparing their enrollment data in Figure 2.4.2.

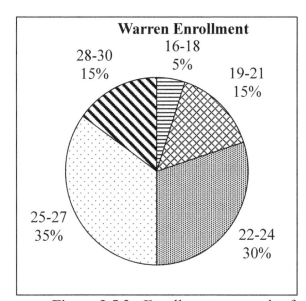
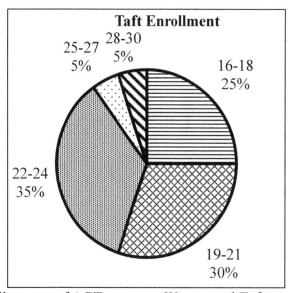

Figure 2.5.2: Enrollment summaries for all ranges of ACT scores at Warren and Taft

The differences in the profiles of each university now look quite stark. Ms. Stats then proceeded to explain how Taft has a lower overall six-year graduation rate. The overall graduation probability is determined by calculating a weighted sum of the individual values. This involves

- partitioning the student population into ACT score groups
- determining the conditional probability of graduating for each group
- calculating a weighted sum

Let G = student graduates within six years and
      R_i = proportion of students in Range i.

The formula for calculating the overall graduation rate is:

$$P(G) = P(R_1) \cdot P(G|R_1) + P(R_2) \cdot P(G|R_2) + P(R_3) \cdot P(G|R_3) + P(R_4) \cdot P(G|R_4) + P(R_5) \cdot P(G|R_5)$$

Using summation notation, this long equation can be written much more concisely.

$$P(G) = \sum_{i=1}^{5} P(R_i) \cdot P(G|R_i)$$

In Warren University there is a 0.45 probability that a student with an ACT score in the range of 16-18 will graduate within six years. The corresponding probability for similar students at Taft University is 50%. However, only 5% of the Warren University students fall within this range. At Taft, 25% of the students had ACT scores in the range of 16-18.

Ms. Stats then proceeded to calculate the overall probability for Warren University using the partition rule for total probability.

$$P(G) = (0.05)(0.45) + (0.15)(0.52) + (0.30)(0.59) + (0.35)(0.65) + (0.15)(0.71)$$
$$\approx 0.61$$

1. She asked them to calculate the same probability for Taft University.

In aggregate Taft University performed poorer because its mix of students tended towards those with lower ACT scores. Ms. Stats told them to focus on the conditional probability of graduation for students in their ACT score range.

2. What is the conditional probability of graduation for each university for students with an ACT score similar to Donald's?

3. What is the conditional probability for each university for students with an ACT score similar to Serena's?

2.5.1 Improve the Graduation Rate

Dr. Miguel Herrera is Vice President for Academic Affairs at Taft University. His university was recently targeted by newspaper stories that highlighted the fact that its graduation rate was below 60%. The university's Board of Governors challenged Dr. Herrera to improve the overall rate to 60%. Dr. Herrera gathered ACT data for last year's incoming class of 500 students. The data are reported in Table 2.4.2.

ACT Scores	Students
16-18	125
19-21	150
22-24	175
25-27	25
28-30	25
Aggregate	500

Table 2.5.2: ACT and enrollment data for Taft University

One suggestion was to increase the minimum standard of admission to the university to 19 on the ACT exam.

4. Would this change achieve the goal of a 60% graduation rate?

Dr. Herrera did not seriously consider this option. The university could not afford to reduce the incoming class by 125 students. On average, each student generated $15,000 in revenue for the school for each year of attendance.

5. How much revenue would be lost if the admission standard was raised to 19 in the first year of this plan?

One option he considered was limiting the number of students in the two lowest categories. He proposed limiting the lowest category to 100 students and the 19-21 category to 125 students.

6. Would this enable the school to achieve a 60% graduation rate?

7. How much revenue would they lose?

Professor R. Prime suggested offering a limited number of merit based scholarships for students with high ACT scores. He proposed offering a $10,000 merit scholarship. These scholarships could attract 25 more students in each of the highest two categories. This would compensate for the caps on students with low ACT scores. This would result in 500 students in the entering class. With these scholarships the net revenue per student for these merit scholars would be only $5,000.

8. Would these scholarships along with the cap on students with low ACT scores enable the school to achieve a 60% graduation rate?

9. How much revenue would they lose?

Dr. Herrera was proud of Taft University's performance with the lowest performing students. However, he recognized there was still room for improvement. The best universities were able to achieve a 55% graduation rate for students in the 16-18 range. Some universities graduated 60% of the students with ACT scores in the 19-21 range. He wanted to evaluate a different alternative. He reviewed several programs at other universities. Dr. Herrera concluded it would cost an additional $1,200 per student in tutoring and mentoring to achieve comparable results for students with the lowest ACT scores. For the following questions, assume no change in admission policy.

10. Would this enable the school to achieve a 60% graduation rate?

11. How much would this cost the school per year?

Section 2.6 Testing for Celiac Disease

Tests of all kinds are used as tools for classifying. The tests you take in school aim to categorize your level of proficiency on a topic. Medical tests often seek to confirm or exclude the presence of a disease. Your email provider likely tests each message you receive to determine if it is spam (i.e., an unsolicited message) and filters it accordingly. Astronomers use tests to measure how likely it is that a given astronomical object is an Earthlike planet. Hearing aids use tests to determine which sounds to amplify and which to disregard. Cars process signals from sensors to decide when conditions are right to deploy air bags. Computer security software tests files and programs in an effort to quarantine anything infected with a virus. An emergency room physician needs to analyze a patient's electrocardiogram to determine if the patient should be admitted to the hospital or sent home. Tests of all kinds are everywhere!

2.6.1 False Positives and False Negatives

No test is perfect, however. A good test is likely to give a positive result when what it is looking for is actually present. It is also unlikely to give a positive result when what it is looking for is not present. But errors can occur, as outlined in Table 2.6.1.

		Object of Interest	
		Present	Not Present
Test Result	Positive	True Positive	False Positive
	Negative	False Negative	True Negative

Table 2.6.1: Possible true and false outcomes of a test

If a test result indicates a given condition when it is actually not present, the test result is said to be a *false positive*. A false positive could occur if a woman performed a home pregnancy test and the test result was positive even though the woman was, in fact, not pregnant. False positive test results can have emotional, safety, and financial ramifications.

1. Describe a situation where you could say your score on a multiple-choice test included a false positive result on a question.

A *false negative* occurs if a test result is negative when it should not be. This would be the case if a test for a concussion came back negative when the patient actually had a concussion. There are, of course, harmful consequences for false negative test results as well. In the case of medical tests, it is particularly important to develop tests with a very small rate of false negatives.

2. Describe a situation where you could say your multiple choice test score constituted a false negative result on a specific question.

Consider mammography, which is a noninvasive test that utilizes low-energy X-rays. This test is widely used to screen for breast cancer in women over 40 years of age. Occasionally a mammogram will show an abnormal lump when there is no cancer present. When patients are notified of the positive test result, they usually become anxious about their prospects for having breast cancer. The mammogram is not a definitive test, though. After the positive mammogram, women are often prescribed a biopsy. Biopsies are more expensive, invasive, and more painful than mammography.

3. Describe some adverse effects of a false negative mammogram.

Knowing how to interpret test results is vitally important because of the consequences of errors. Interpreting test results requires an understanding of conditional probability. For example, consider a patient who has blood drawn at a laboratory to screen for HIV. When the report for that blood test goes back to the patient's physician with a positive result, the physician needs to answer the question, "What is the probability the patient actually has HIV, given the positive test result?" The answer depends on the rates of false positives and false negatives of the test as well as demographic information about the patient (e.g., age, gender, race, history of sexual activity and drug use). Medical tests have well documented rates of giving false positive and false negative results. Experts on diseases are also able to identify risk factors (genetic or environmental) that increase a person's likelihood of having a given disease.

As you have seen, conditional probability is not something we human beings have good intuition about. Even highly educated and well trained professionals have difficulty in these situations. To illustrate this point, consider a study conducted by German cognitive psychologist Gerd Gigerenzer in 2002. He gave the following scenario to 100 American doctors.

> *The probability that one of these women has breast cancer is 0.8 percent. If a woman has breast cancer, the probability is 90 percent that she will have a positive mammogram. If a woman does not have breast cancer, the probability is 7 percent that she will still have a positive mammogram. Imagine a woman who has a positive mammogram. What is the probability that she actually has breast cancer?*

4. What is your estimate of the likelihood this woman has breast cancer?

Of the 100 doctors studied, 95 of them estimated the woman's probability of having breast cancer to be around 75%. The correct probability, given the provided information, is about 9%.

The correct answer is so low due to the fact that breast cancer is relatively rare at any particular time in the life of a woman. It was stated in the problem that breast cancer affects only 0.8% of women in this population. Imagine a group of 1,000 women, all of whom are going to be screened for breast cancer. On average, only eight of the 1,000 women will actually have breast cancer. Many more than that will receive false positive tests, simply because the vast majority of these women do not have breast cancer. The 0.8% rate of women having breast cancer is called the ***prior probability*** and plays a major role in interpreting test results, as you will soon see.

2.6.2 Diagnosing Celiac Disease

Lily has not been feeling well for the past few months. Her symptoms are generally restricted to her digestive system. They include abdominal swelling, frequent diarrhea, and occasional vomiting. She is not at all sure of what is causing these symptoms. However, she sees the wide variety of gluten-free foods on the market and wonders if she might be gluten intolerant.

Celiac disease is a chronic (i.e., lifelong) autoimmune disease that causes inflammation in the small intestine. Autoimmune diseases are diseases in which one's immune system attacks one's own tissues and cells. In the case of celiac disease, antibodies attack the villi that line the small intestine. Villi are tiny finger-like structures which increase the surface area within the small intestine in order to absorb more nutrients from food. (See Figure 2.6.1.) When people with celiac disease eat gluten, their villi flatten out as their intestinal lining is damaged by antibodies. This leads to a decrease in the person's ability to absorb nutrients.

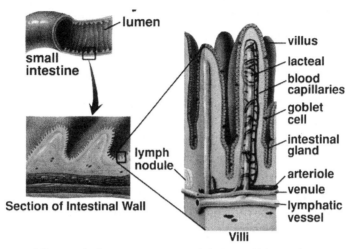

Figure 2.6.1: Anatomy of the small intestine

The **_prevalence_** of celiac disease in the United States is approximately 1%. This means that in any random sample of 100 Americans, on average, one person will have celiac disease. The prevalence of a disease shows the proportion of a population who have the disease. It can also be interpreted as the probability that a randomly selected person in the population has the disease.

Lily's physician, Dr. Grano, orders a blood test to check for the presence of a particular antibody, anti-tissue transglutaminase immunoglobulin A (tTG-IgA). This is a biomarker of celiac disease. Medical tests often are unable to directly detect a disease, so they check for biomarkers instead. Biomarkers are other cells, molecules, or genes that are associated with the disease and can be detected and measured more easily. However, the relationship between a biomarker and a disease is not perfect. All medical tests will occasionally give false results. They report either the existence of a disease that is not present (a false positive) or fail to recognize a disease that is present (a false negative). Sensitivity and specificity are statistical measures that describe how well medical tests work. The **_sensitivity_** of a test is the probability of the test showing a positive result when the disease is actually present. Another way of saying that is that the sensitivity of a test is the rate of true positive results from the test. The **_specificity_** of a test is the probability of the test showing a negative result when the disease is not present. Again, this could also be called the true negative rate of the test.

According to the National Institutes of Health, the sensitivity of the tTG-IgA test is 93% and its specificity is 98%. The tTG-IgA blood test has two possible outcomes: positive or negative.

The sensitivity and specificity of tests can be used to calculate the rates of false test results, as shown below.

False positive rate = 1 – specificity

False negative rate = 1 – sensitivity

The rate of false positive test results could also be explained as a conditional probability. Specifically, the false positive rate is the probability of getting a positive (+) test result given no disease is present.

False positive rate = $P(+\text{ test} \mid \text{no celiac})$

5. Write the conditional probability associated with each rate.
    a. True positive rate
    b. True negative rate
    c. False negative rate

6. The sensitivity of a test indicates the rate of true positive results. Explain why the false positive rate does not equal one minus the sensitivity of the test.

Dr. Grano explains the pathology and physiology of celiac disease to Lily. Before ordering the blood test, Dr. Grano also counsels Lily on the strengths and limitations of the tTG-IgA blood test. Lily hears the 93% sensitivity and 98% specificity; she feels very confident in the ability of the test to confirm or disconfirm her self-diagnosis of celiac disease.

Several days later, the results of Lily's blood work comes back positive for tTG-IgA. Dr. Grano interviews Lily to determine if there is a family history of celiac disease. Dr. Grano also asks if Lily has experienced any fatigue or weight loss that coincided with her other symptoms. Lily reported that there is no family history of celiac disease. She also indicated that she did not experience the fatigue or weight loss that are common to celiac disease.

If Dr. Grano endorsed the celiac disease diagnosis, her next step would be to prescribe a treatment plan. The typical treatment for celiac disease is a gluten-free diet. This lifestyle change would mean a radical shift in how Lily currently eats. Before making such a drastic recommendation, Dr. Grano considers ordering a second test more reliable to confirm the diagnosis.

The "gold standard" test for diagnosing celiac disease is an upper gastrointestinal endoscopy. An endoscope is a thin scope with a light and camera at its tip. To check for celiac disease, the endoscope is inserted through the mouth and guided through esophagus and the stomach until entering the beginning of the small intestine. (See Figure 2.6.2.)

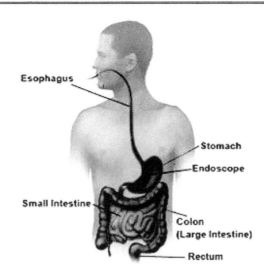

Figure 2.6.2: Digestive system with endoscope

If abnormal tissue is observed during the endoscopy, it can be sampled and sent to a lab for biopsy. In the biopsy, a pathologist will examine the abnormal tissue under a microscope to determine if the patient actually has celiac disease. This is the gold standard test because it is the best test available to diagnose celiac disease.

7. Give some reasons why the "gold standard" test would not be the first test used to diagnose a disease.

Dr. Grano advises Lily to undergo the upper gastrointestinal endoscopy. Lily listens patiently as Dr. Grano explains how to prepare for the test and what will occur during the procedure. The preparation includes not eating anything the day of the procedure and not drinking anything for the final four hours before the endoscopy. This ensures that the stomach will be empty. This will allow the endoscope to get a clear picture from within Lily's small intestine. She will also have to be sedated during the procedure.

Lily is not thrilled at the prospect of having to undergo this expensive, invasive, and inconvenient test. She is confident in the blood test and does not see the need for the biopsy.

Dr. Grano knows that the vast majority of people have poor intuition regarding probability in general and conditional probability in particular. The relevant probabilities are shown in Figure 2.6.3. These numbers are abstract and difficult to interpret meaningfully.

Let T = positive test result
 T^c = negative test result
 C = has celiac disease
 C^c = does not have celiac disease

The four possible combinations of test results and disease can be represented with both a probability tree, Figure 2.6.3, and a table, Table 2.6.2. In the tree, a path corresponds to a pair of events. At the origin node, there are two possible branches that represent the random event whether or not the individual has the disease. This leads to a second random event, the test results. The test results can either be positive or negative. On each branch is placed the probability of that event. For example, the topmost path begins with the probability of having celiac. This probability is 0.01. Next is the conditional probability of obtaining a positive test result given the individual has the disease. This probability is 0.93. The probability at the end of the path represents the joint probability, $P(C \cap T)$. It is calculated by using the formula

$$P(C \cap T) = P(C) \cdot P(T|C)$$
$$= (0.01)(0.93) \quad \text{or} \quad \text{about 1 in 108}$$
$$= 0.0093$$

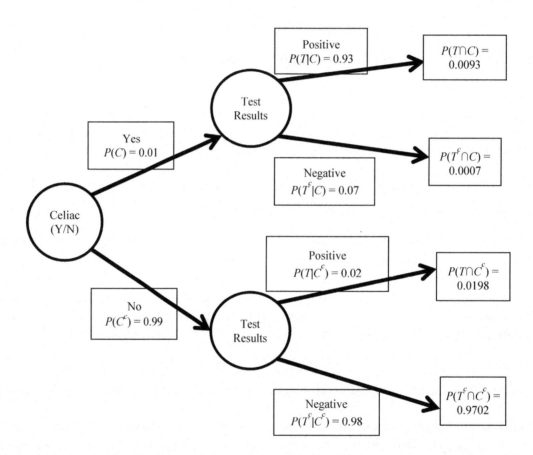

Figure 2.6.3: Tree diagram showing probabilities of possible outcomes for celiac disease test

Instead of focusing on all the probabilities, Dr. Grano tries to justify her recommendation by asking Lily to consider a sample of 10,000 people. The prevalence of celiac disease in the US is 1%. Therefore, on average, 100 people in the sample of 10,000 will have celiac disease. The sensitivity of the tTG-IgA test is 93%. This means that on average 93% of the 100 people who have celiac disease will test positive for it. Thus, there will be seven people who have the disease but will get a negative test result.

If only 100 of the 10,000 people actually have celiac, then 9,900 of them do not have the disease. The specificity of the tTG-IgA test is 98%. This means that 98% of the 9,900 people without celiac will receive negative test results. That leaves 198 people who do not have celiac but receive positive test results. The results of this discussion are summarized in Table 2.6.2. Each of the four center cells represents the intersection of two events. For example, there are 9,702 people who do not have celiac and test negative out of the 10,000. Thus, the probability that a randomly selected individual who is tested will not have the disease and test negative is 0.9702. If we want to calculate the conditional probability within a row, we divide by the row total. There were a total of 291 people with positive test results. Out of this total, 93 have the disease. This discussion leads to the following conclusion. Given that Lily received positive test results, the probability she has the disease is (93/291) or 0.32. For most people this low probability is counterintuitive.

		True condition		Totals
		Celiac	No celiac	
tTG-IgA test result	Positive test	93 True positive $T \cap C$	198 False positive $T \cap C^c$	291 Positive test results
	Negative test	7 False negative $T^c \cap C$	9702 True negative $T^c \cap C^c$	9,709 Negative test results
Totals		100 Celiac	9,900 No celiac	

Table 2.6.2: Average outcomes for a sample of 1,000 people being tested for celiac disease

8. Given these calculations, do you think it is reasonable for Lily to undergo the upper gastrointestinal endoscopy procedure to confirm the celiac disease diagnosis before changing her diet?

The test is a much better predictor if the test results had been negative.

9. How many negative test results on average occur in the hypothetical sample of 10,000 people?

10. How many of the negative test results were true negatives?

11. Based on the above information, what would be the probability that Lily did not have celiac disease, if she got a negative tTG-IgA test result?

The impact of a false negative is more consequential than the impact of a false positive. A false negative means that the patient's illness was not discovered. In contrast, a false positive often entails additional tests. These tests are usually much more expensive and invasive. In addition, patients worry about their future until receiving the more accurate test results. Recently, the medical community has begun to review their recommendations on annual testing for relatively rare ailments with significant false positive rates.

Section 2.7 Bayes Theorem

The methodology used above to calculate the conditional probabilities is related to Bayes Theorem. The theorem was developed by Reverend Thomas Bayes in the 18th Century. As has been noted above, the probability of the intersection of two events, *A* and *B*, can be calculated using conditional probability. However, it is possible to condition on either of the two events, *A* or *B*. The two representations are thus equivalent.

$$\left. \begin{array}{l} P(A \cap B) = P(A) \cdot P(B|A) \\ P(A \cap B) = P(B) \cdot P(A|B) \end{array} \right\} \rightarrow P(A) \cdot P(B|A) = P(B) \cdot P(A|B)$$

From the above equation it is possible to solve for one conditional probability in terms of the other.

$$P(B|A) = \frac{P(B) \cdot P(A|B)}{P(A)}$$

One way to calculate a probability of an event is to decompose it into all of the possible ways it can occur with a second event. This involves partitioning the second event's sample space. The simplest partition of an event space involves just two complementary events, *B* and B^c. Using the concept of a partition to calculate a total probability, the denominator of the above equation is equal to the following.

$$P(A) = P(A \cap B) + P(A \cap B^c)$$
$$= P(B) \cdot P(A|B) + P(B^c) \cdot P(A|B^c)$$

When we substitute for *P(A)* in the denominator, the original equation becomes

$$P(B|A) = \frac{P(A \cap B)}{P(A \cap B) + P(A \cap B^c)}.$$

Alternatively,

$$P(B|A) = \frac{P(B) \cdot P(A|B)}{P(B) \cdot P(A|B) + P(B^c) \cdot P(A|B^c)}.$$

Notice that the numerator term is also one of the terms in the denominator.

		True condition		Test Result Partition
		Celiac	No celiac	
tTG-IgA test result	Positive test	True positive $P(T \cap C) = \dfrac{93}{10,000}$	False positive $P(T \cap C^c) = \dfrac{198}{10,000}$	$P(T) = \dfrac{291}{10,000}$
	Negative test	False negative $P(T^c \cap C) = \dfrac{7}{10,000}$	True negative $P(T^c \cap C^c) = \dfrac{9,702}{10,000}$	$P(T^c) = \dfrac{9,709}{10,000}$
Disease Partition		$P(C) = \dfrac{100}{10,000}$	$P(C^c) = \dfrac{9,900}{10,000}$	

Table 2.7.1: Probability of different outcomes for people being tested for celiac disease

Let's revisit Table 2.6.2 and convert the number of people to probabilities. In the Celiac example, the conditional probabilities are determined by taking a specific cell value and dividing by either the row or column total. For example, we divide by the column total, probability of event C, to obtain P(T|C).

$$P(T \mid C) = \frac{P(T \cap C)}{P(T \cap C) + P(T \cap C^c)}$$

$$= \frac{\frac{93}{10,000}}{\frac{100}{10,000}}$$

$$= 0.93$$

This probability corresponds to the 93% sensitivity of the test that was noted in the original description of the test's accuracy.

We divide by the row total, probability of event T, to determine P(C|T). Given that the test results were positive, the chance of actually having the disease is 0.32.

$$P(C \mid T) = \frac{P(C \cap T)}{P(C \cap T) + P(C \cap T^c)}$$

$$= \frac{\frac{93}{10,000}}{\frac{291}{10,000}}$$

$$\approx 0.32 \qquad \text{or} \qquad \text{about 1 in 3}$$

This also means that approximately 68% of people who receive positive results do not have celiac disease.

Chapter 2 (Conditional probability) Homework

1. The pairs of events below could be assumed to be independent or dependent under different circumstances. Describe a situation in which the events would be independent and another situation in which the events would be dependent.

    a. The passenger in seat 12A misses a flight and the passenger in seat 12C misses the same flight.

    Independent if:

    Dependent if:

    b. A male dying from cancer and a female dying from cancer.

    Independent if:

    Dependent if:

    c. The mortgage on one house going into foreclosure and the mortgage of a house three miles away going into foreclosure.

    Independent if:

    Dependent if:

    d. You being absent from school and a classmate being absent from school.

    Independent if:

    Dependent if:

    e. Getting an *A* on a test and getting a *C* on the next test.

    Independent if:

    Dependent if:

2. The data below is an excerpt from the Social Security Administration's actuarial life table. It gives the probability that a randomly selected individual will die within a particular year of life. It also shows the number of individuals in a theoretical sample of 100,000 who would survive that year. These data are shown for males and females.

Exact age	Male		Female	
	Probability of dying within one year	Number of survivors out of 100,000 born alive	Probability of dying within one year	Number of survivors out of 100,000 born alive
0	0.00668	100,000	0.005562	100,000
1	0.000436	99,332	0.000396	99,444
2	0.000304	99,289	0.000214	99,404
3	0.000232	99,259	0.000162	99,383
4	0.000172	99,235	0.000132	99,367
5	0.000155	99,218	0.000117	99,354
6	0.000143	99,203	0.000106	99,342
7	0.000131	99,189	0.000099	99,332
8	0.000115	99,176	0.000093	99,322
9	0.000096	99,164	0.00009	99,313
10	0.000082	99,155	0.00009	99,304
11	0.000086	99,147	0.000096	99,295
12	0.000125	99,138	0.000111	99,285
13	0.000205	99,126	0.000137	99,274
14	0.000319	99,106	0.00017	99,261
15	0.000441	99,074	0.000207	99,244
16	0.000562	99,030	0.000245	99,223
17	0.00069	98,975	0.000282	99,199
18	0.00082	98,906	0.000318	99,171
19	0.000949	98,825	0.000352	99,139
20	0.001085	98,731	0.000388	99,105
21	0.001213	98,624	0.000423	99,066
22	0.001304	98,505	0.000454	99,024
23	0.001345	98,376	0.000476	98,979
24	0.00135	98,244	0.000494	98,932
25	0.001342	98,111	0.000511	98,883
26	0.00134	97,980	0.000531	98,833
27	0.001342	97,848	0.000553	98,780
28	0.001356	97,717	0.000579	98,726
29	0.00138	97,584	0.000608	98,668

Table 1: Mortality tables through age 29

a. In the original sample of 100,000 male lives, on average, how many individuals will survive to age 25?
b. Find *P*(dying before age 28 | male).
c. Given that a male from this group survived to age 22, how likely is he NOT to live to see his 28th birthday?
d. Find the probability that a female who is alive at age 18 will die before reaching her 29th birthday.
e. What percentage of the original 100,000 females is living at the age of 20?
f. What is probability that a randomly selected (living) 13-year-old girl will die before she turns 25?
g. How likely is it that an 18-year-old male will die within the next four years?

3. The names of nine students are on separate slips of paper and placed in a hat. One of the names is yours. Two names will be drawn at random to represent your school.

   a. What is the probability that your name will be drawn?
   b. Consider the case where the slips of paper with names on them were placed into two hats: one hat contains all the clever students (of which there are five) and the other contains the dim students. One student will be chosen from each hat. You, of course, are clever. What is the probability that your name will be drawn?

4. Central Heights High School (CHHS) is participating in a program sponsored by the American Civil Liberties Union. The goal of the event is to raise awareness of issues of racial profiling in local police work. Nine students, six black and three white, have applied to participate in the program, but each school is only allowed to send two participants. CHHS's administration would like to send a racially balanced group but does not necessarily want to give the perception that it is using a quota system for selecting the participants.

   a. If selecting students to participate is a random process, what is the probability the first student selected is black?
   b. If the first randomly selected student is black, what is the probability the second student selected will also be black?
   c. If both participants are selected randomly, what is the probability that one black and one white student will be chosen?
   d. What is the probability that both randomly selected participants are white?

5. New Westshire High School just completed its 8th annual charity month fundraising campaign. Through 50-50 raffles, car washes, bake sales (of USDA compliant snacks, of course), and a host of other events, the school has raised nearly $12,000. Now the task of distributing these funds to worthy charities falls to the Student Congress. The New Westshire School Board has approved a list of 14 charities to which the Student Congress is authorized to donate. These include five local charities, and nine national charities. After spending hours arguing over which charities are most worthy of their donations, the students agree to randomly select which charities will get their funds. Student Congress will evenly split the money between four charities.

a. What is the probability that no local charities will receive donations?
b. What is the probability that the selected charities will include a mix of local and national organizations?
c. If they decide to only donate to three charities, what is the probability that all three will be local?

6. Daniella Kidman is an outstanding rebounder for the Dukes. However, she is one of the weaker foul shooters on the team. The data in Table 2 records 30 pairs of shots Daniella took over the last ten games. An X means she made the free throw; an O means she missed the free throw.

Pair	Shot 1	Shot 2	Points	Pair	Shot 1	Shot 2	Points	Pair	Shot 1	Shot 2	Points
1	X	O	1	11	X	X	2	21	X	X	2
2	O	X	1	12	O	O	0	22	X	O	1
3	X	X	2	13	X	X	2	23	X	X	2
4	X	X	2	14	O	O	0	24	O	X	1
5	O	O	0	15	X	X	2	25	O	O	0
6	X	O	1	16	X	X	2	26	O	O	0
7	O	O	0	17	O	O	0	27	X	O	1
8	X	X	2	18	X	X	2	28	X	X	2
9	X	X	2	19	O	O	0	29	X	O	1
10	O	X	1	20	X	X	2	30	X	X	2

Table 2: Kidman's foul shooting

a. What is the probability that she makes a foul shot?
Coach Blu wondered if Daniella did about as well on the first shot as on the second shot.
b. What is the probability that she makes her first foul shot?
c. What is the probability that she makes her second foul shot?
In any data sample there is random fluctuation.
d. Explain whether or not you think the different probabilities for the first and second shot might simply be random fluctuations around her rate of sinking free throws.

7. Coach Blu did notice an interesting pattern. When Daniella made the first free throw, she seemed to relax. When she missed her first free throw she seemed to tense up. She believes that this might affect her performance on the second free throw.

a. What is the conditional probability that she sinks the second free throw, after she has made the first free throw?
b. What is the conditional probability that she sinks the second free throw, after she has missed the first free throw?

c. Do the probabilities provide evidence to support or disprove Coach Blu's observations?

Coach Blu is interested in the probability distribution of the number of points Daniella makes when she is fouled in the act of shooting.

d. Complete the table below

y	$P(Y=y)$
0	
1	
2	

e. What is the expected value, $E(Y)$, of the number of points she scores when taking two free throws.

8. Coach Blu is especially concerned at what happens towards the end of the half. At that point, the team is in a 1 and 1 situation. If Daniella is fouled, she takes a second shot only if she makes the first. Over the last ten games, this situation has not happened very often. Coach Blu comes up with idea of using the original data set in Table 3. Then imagining what the results would be if it were a one-and-one situation. For example, the first pair Daniella made the first shot. Therefore, it would be unchanged. However, with the second pair, Daniella missed the first shot. She would therefore not have a second shot.

Pair	Shot 1	Shot 2	Points	Pair	Shot 1	Shot 2	Points	Pair	Shot 1	Shot 2	Points
1	X	O		11	X	X		21	X	X	
2	O			12	O			22	X	O	
3	X	X		13	X	X		23	X	X	
4	X	X		14	O			24	O		
5	O			15	X	X		25	O		
6	X	O		16	X	X		26	O		
7	O			17	O			27	X	O	
8	X	X		18	X	X		28	X	X	
9	X	X		19	O			29	X	O	
10	O			20	X	X		30	X	X	

Table 3: Daniella's foul shooting at end of a half

a. Complete the table below

y	$P(Y=y)$
0	
1	
2	

b. What is the expected value, $E(Y)$, of the number of points she scores when taking two free throws?

9. Mr. I. M. Bother was concerned about the amount of spam he was reading. He reviewed each of his most recent 200 e-mails. He classified each message as spam or not spam. The data are summarized in the Table 4. A total of 120 messages were sent to his Inbox. His spam filter treated them as not Spam. He had a total of 80 messages in his Spam box.

	Actual		
	Not Spam	Spam	Totals
Inbox	80	40	120
Spam Box	10	70	80
Totals	90	110	200

Table 4: Spam data

a. What is the probability that a randomly selected e-mail sent to Mr. Bother was spam?
b. What is the probability that a randomly selected e-mail in his Inbox is spam?

A spam e-mail can end up either in the Inbox or the spam box.

c. What is the probability that a spam e-mail ends up in his spam box?

Mr. Bother is also concerned about missing e-mails that are not spam.

d. What is the probability that a non-spam message is sent to the spam box.

Mr. Bother's school was just attacked with massive amounts of spam. As a result 90% of the messages coming to him are spam.

e. Now, what is the probability that a randomly selected e-mail in his Inbox is spam?

Until this attack is over, Mr. Bother plans on resetting his spam filter to keep out more spam. With his new settings, the probability that a spam e-mail is filtered out and sent to the spam box is 0.95. However, this change also increases the likelihood that non-spam also is sent to the spam box. With this setting, there is a 0.25 probability that a non-spam e-mail is sent to the spam box.

f. What is the probability that a randomly selected e-mail in his Inbox is still spam?

10. A traffic accident occurred one dark night in Gotham City.[2] Luckily, Mr. I.M. Schure witnessed the accident. Mr. Schure testified that he saw a blue taxi swerve and hit

---

[2] A. Tversky, D. Kahneman, "Evidential impact of base rates", in ***Judgement under uncertainty: Heuristics and biases***, D. Kahneman, P. Slovic, A. Tversky (editors), Cambridge University Press, 1982.

another car; it was the blue taxi's fault. In Gotham there are only two taxi companies. Blue Bat Taxi Company has blue taxis that make up 25% of the total taxis in the city. Lucky Leprechaun Taxi Company has green taxis and account for 75% of the taxis in the city. The accident rate per vehicle is approximately the same for both taxicab companies. I.M. Schure was shown 100 taxis, fifty of each company. He was asked to identify their color under similar night time conditions. He correctly identified 90% of the blue taxis as blue. He correctly identified 80% of the green taxis as green.

   a. What is the probability that a taxicab involved accident involved a Blue Bat taxicab?
   b. What is the probability that a witness of a random taxicab accident, would claim he saw a blue taxicab?
   c. Given Mr. Schure's testimony, what is the probability that the accident involved a Blue Bat taxi?
   d. Do you think Mr. Schure's testimony is enough evidence to hold Blue Bat Taxi Company liable for this accident? Explain.

11. The Beltway Insurance Company is considering a plan to use the first three years of driving history to identify and divide their customers into two categories: high risk and low risk.[3] Among their customers, high risk drivers have twice as many accidents as low risk drivers. Beltway estimates that 25% of its drivers are high risk. These drivers have of 0.3 probability of having a car accident during a year. Low risk drivers are estimated to have a 0.15 probability of having an accident.

   a. What is the probability that a randomly selected driver will have an accident during the year?
   b. Daniel Lion had an accident this past year. What is the probability that he is a high risk driver?
   c. What is the probability that a randomly selected driver will have an accident two years in a row?
   d. Jeanette Kelly had an accident two years in a row. What is the probability that she is a high risk driver? Is there enough evidence to charge her a much higher insurance premium for being a high risk driver?
   e. What is the probability that a randomly selected driver will have an accident three years in a row?
   f. Jerry Vino had an accident three years in a row. What is the probability that he is a high risk driver? Is there enough evidence to charge him a much higher insurance premium for being a high risk driver?

In a densely populated area, accidents rates are higher. Assume that in Gotham City, a low risk driver has a 0.2 probability of having an accident. The corresponding probability for a high risk driver is double that or 0.4.

   g. Repeat the analysis of parts a through f. What do you notice about the results?

---

[3] This example is adapted from Sheldon Ross, *A First Course in Probability*, (Pearson).

12. There are 23 pairs of chromosomes in the cell nuclei of healthy humans. Each of these 23 pairs is uniquely identified and numbered. One class of genetic defects is named trisomes. This defect means the individual has three chromosomes instead of two in a specific location. Down Syndrome, also called trisomy 21, has an extra chromosome on the 21st pair. Trisomy 18 (Edwards Syndrome) and trisomy 13 (Patau Syndrome) are two other syndromes. These two anomalies are much less common that Down but have a far greater impact on human development and early death. The majority of fetuses with either of these two syndromes will not be born alive either as a result of spontaneous miscarriage or stillbirth. For those who survive birth, the median lifespan is less than 10 days and only an estimated 10% survive a year or more.[4]

In 2011 a new genetic test became available to screen for some of these conditions. The test does not require the invasive procedure of extracting amniotic fluid surrounding the fetus. Researchers found that fragments of fetal DNA float freely in the blood of a pregnant woman as early as seven weeks into the pregnancy. The amount increases as the pregnancy proceeds. These free floating fragments of fetal DNA can be identified and probed for genetic defects with highly accurate test methodologies. The test sensitivity is the likelihood of detecting the defect in a fetus with the genetic disorder. The test specificity is the likelihood of a negative diagnosis when the fetus does not have the genetic defect.

- test sensitivity → P(positive diagnosis | trisomy 21) = 0.993
- test specificity → P(negative diagnosis | no trisomy 21) = 0.998

a. What is the likelihood of a false positive result with a fetus that does not have the genetic defect?

The likelihood that a fetus carries one of these defects grows significantly with the age of the mother along with other factors such as a positive family history for these conditions. Assume, for example, that the prevalence of the trisomy 21 defect is 1 in 1,000 for a mother aged 25 years without other identified risks factors.

b. Assume 10,000 similar pregnant women who were 25 years old were tested. On average how many fetuses would have the genetic defect and how many would not? (Round to the nearest integer.)
c. On average how many true positive results will be found in this population of 10,000?
d. On average how many false positive results will be found in this population of 10,000?

A pregnant woman just received a report of a positive result using this new screening test. Her genetic counselor sought to clarify for her that it does not mean the fetus definitely has the disorder.

---

[4] Morris JK, and Savva GM. (2008) The risk of fetal loss following a prenatal diagnosis of trisomy 13 or trisomy 18. *Am J Med Genet* Apr 1:146A(7):827-32.

e. For this 25 year old woman, use the above averages to calculate the probability that the fetus has the trisomy 21 defect.
f. Also apply Bayes rule to calculate this probability.

13. Now assume the pregnant woman is 40 years old and the prevalence of the trisomy 21 defect is 1 in 75 for women that age.

    a. Calculate P(trisomy 21| positive diagnosis) for a 40-year-old woman.
    b. Describe the significance of knowing the prevalence of a genetic defect when interpreting the test results.

Expanded Homework Problems

I. Switch hitting batting average

1. Ted Rubio is a switch hitting second baseman for the Washington Senators baseball team. When facing right-handed pitching, he bats left-handed. Conversely, he bats right-handed when facing a left-handed pitcher. Table 5 contains his statistics for the first half of the season. When batting right-handed, his probability of getting a hit was almost one in three. When batting left-handed, the probability was only 0.25. Overall, his chance of getting a hit during the first half of the season was slightly more than three in ten. His batting average was .302. Ted was not satisfied with his performance batting left-handed. He sought guidance from Marc Cruz, the Senator's new hitting coach. Coach Cruz offered Rubio his guidance during extra left-handed batting practice. He made small adjustments in Ted's batting stance.

	Right-Handed	Left-Handed	Total
Hits	53	23	76
At-bats	160	92	252
Batting Average	.331	.250	.302

Table 5: Batting averages first half of season

During the second half of the season, Ted felt more comfortable batting left-handed and his average improved to .272. He had 49 hits in 180 left-handed at-bats as reported in Table 6. However, when Ted looked at his total performance in the last 81 games, he was disappointed. His overall batting average had declined in the second half of the season. His batting average had fallen below .300. He wondered if extra practice he put into batting left-handed, had hurt his performance while batting right-handed.

	Right-Handed	Left-Handed	Total
Hits	32	49	81
At-bats	97	180	277
Batting Average	.330	.272	.292

Table 6: Batting averages second half of season

The manager for the Senators, Harry McConnell, had a master's degree in statistics. He pointed out that Ted's right-handed batting average in the second half of the season was almost identical to his performance earlier in the year. It was .330 as compared to the previous .331. That was not the explanation for the ten point drop in overall average. He then proceeded to try to explain the drop in the overall batting average by using the concept of a partition.

    a. During the first half of the season what proportion of at-bats were right-handed and what proportion were left-handed?
    b. Use this information and the concept of a partition to show another way to calculate his batting average for the first half of the season.
    c. During the second half of the season what proportion of at-bats were right-handed and what proportion were left-handed?
    d. Use this information and the concept of a partition to show another way to calculate his batting average for the second half of the season.
    e. Explain why his overall second half batting average was lower even though he improved his left-handed hitting.
    f. What would his second half overall batting average have been if the second half percentage of left-handed at-bats were the same as the first half percentage?

II. Heart valve surgery

1. Mr. Robert Hart was asked by his 70-year-old mother Alison to help her decide where to have surgery. She was diagnosed with serious heart valve problems and was told she would need surgery to address the problem. She understood it was risky surgery but did not know how risky. Robert reviewed the data reported online for his local area hospital. In 2011, cardiac surgeons at Lonely Heart Hospital had carried out 233 valve replacement surgeries. Of this total, 25 patients died within thirty days of surgery.
    a. What is the likelihood that a randomly selected patient admitted to Lonely Heart Hospital will not survive more than 30 days after valve replacement surgery?

His mother does not yet have a complete assessment of the required surgery. She has some problems with both the aortic valve and the mitral valve. He asked his friend David Analytics, the hospital's chief statistician for more details. Table 7 presents the mortality data for the specific valve replacement alternatives.

Lonely Heart Hospital	2011 Heart Valve Replacement Surgery			
	Aortic	Mitral	Both Valves	Totals
Patients	133	60	40	233
Deaths within 30 days	11	8	6	25

Table 7: Lonely Heart Hospital mortality rates for heart valve surgery in 2011

    b. Which surgical procedure had the lowest risk? Which had the highest risk?
    c. What percentage of valve replacement patients had both valves replaced in 2011?

David also told Robert that he was completing a report for 2012 as shown in Table 8. The hospital had carried out fewer surgeries, a total of 230. However, the number of deaths had increased to 27. The 30 day mortality rate had increased to 11.7% as compared to 10.7% in 2011.

Lonely Heart Hospital	2012 Heart Valve Replacement Surgery			
	Aortic	Mitral	Both Valves	Totals
Patients	52	83	95	230
Deaths within 30 days	4	10	13	27

Table 8: Lonely Heart Hospital mortality rates for heart valve surgery in 2012

Robert expressed concern when he saw the increase in both fatalities and the rate. David told his friend that the data actually told a different story. In fact, Lonely Heart's performance had been better in 2012 than in 2011. David asked Robert to look closely at Table 9 which summarizes the patient mix of surgeries.

Lonely Heart Hospital	Mix of surgeries		
	Aortic	Mitral	Both Valves
2011	133/233 = 57.1%	60/233 = 25.8%	40/233 = 17.2%
2012	52/230 = 22.6%	83/230 = 36.1%	95/230 = 41.3%

Table 9: Lonely Heart Hospital: mix of heart valve surgery in 2011 and 2012

2. Robert noticed that the 2012 mix looked significantly different from the pattern in 2011. In 2011, aortic valve replacement accounted for over 57% of the valve replacement surgeries. That percentage decreased to 23% in 2012. In contrast, both valves were replaced in 17% of the patients in 2011 and more than 41% of the patients in 2012. David then reminded Robert that aortic valve replacement was the lowest risk surgery and double valve replacement was the highest risk. Thus, the mix of surgeries in 2012 was more risky in aggregate when compared to 2011.

David explained why this change in patient mix had occurred. Late in 2011, Lonely Heart Hospital had recruited a nationally renowned cardiac surgeon. They had widely publicized this fact to the general practitioners in the area. Lonely Heart had also let the smaller hospitals in the area know about the changes. They encouraged these physicians and hospitals to send the more complicated heart patients to Lonely Heart Hospital. They had also decided not to aggressively compete for the less serious cardiac surgeries.

David had claimed the performance in 2012 was, in fact, better than in 2011. It was important to look at the conditional probability associated with each procedure

   Let D_1 = the event the patient died in 2011
   A_1 = the event the patient had aortic valve surgically replaced in 2011
   M_1 = the event the patient had mitral valve surgically replaced in 2011
   B_1 = the event the patient had both valves surgically replaced in 2011

   Let D_2 = the event the patient died in 2012
   A_2 = the event the patient had aortic valve surgically replaced in 2012
   M_2 = the event the patient had mitral valve surgically replaced in 2012

B_2 = the event the patient had both valves surgically replaced in 2012

a. In 2011 what was the conditional probability of death with aortic valve replacement?
b. In 2012 what was the conditional probability of death with aortic valve replacement? Did performance get better or worse?
c. Use conditional probability to determine whether or not the probability of dying from mitral valve replacement surgery had improved from 2011 to 2012.
d. Use conditional probability to determine whether or not the probability of dying from double valve replacement surgery had improved in 2012.
e. Use the concept of a partition to determine the overall probability of a cardiac valve replacement patient dying in 2011 and 2012 at Lonely Heart Hospital.

The term $P(A_i)$ represents the proportion of patients in each year i who had aortic valve replacement. In 2011 this proportion was 0.57; in 2012 it was only 0.23

Robert still had trouble understanding the results. How could Lonely Heart Hospital reduce the mortality rate within each valve replacement category but the overall mortality rate increase from 10.7% to 11.7%. David explained that the improvement would be clear if the patient mix had not changed. Imagine the following scenario.
- Surgery specific probability of dying based on 2012 data
- Surgery patient mix the same as in 2011

f. What would the overall probability of dying in 2012 have been if the patient mix had remained the same but performance for each type of surgery had improved as shown by the data

3. Earlier that week, Dr. Fisher had suggested to Robert that he take his mother for an evaluation at Sacred Heart Hospital. Sacred Heart was in an urban center that was 75 miles away. It is a nationally recognized surgical center. Robert had decided not to pursue the suggestion after seeing that their mortality rate was about the same as Lonely Heart. Over a 4 year period, 2009-2012, 728 out of 5800 valve replacement had died within 30 days. The probability of death was 0.126 which is slightly worse than Lonely Heart. However, after learning about conditional probability, he decided to look at the more detailed data by procedure as displayed in Table 4.

4.

Sacred Heart Hospital	Heart Valve Replacement Surgery			
	Aortic	Mitral	Both Valves	Totals
Patients	800	1,500	3,500	5,800
Deaths within 30 days	58	220	450	728

Table 10: Sacred Heart Hospital mortality rates for heart valve surgery 2009-2012

a. Determine the conditional probabilities of death for each type of valve surgery.
b. What advice would you give Alison Hart as to where she should have heart valve replacement surgery?
c. Why do you think Sacred Heart has such a large proportion of double valve surgery?

III. Screening for Tuberculosis (TB) with Mantoux Skin Test

Mycobacterium tuberculosis is the bacterium that causes tuberculosis (TB). The World Health Organization estimates that one-third of the world's population is infected with this bacterium. The vast majority have a latent TB infection. In this stage the individual has no symptoms and is not contagious. However, according to the Center for Disease Control (CDC) he is at 5% - 10% risk of developing the actual disease unless the latent TB is treated. Those that have the TB disease exhibit a number of symptoms. Most of the symptoms are directly related to the respiratory system. Other broader symptoms include fever, chills, fatigue, etc. In the US, approximately three per 10,000 people develop TB disease each year. However, one study estimated that 4% of the US population has latent TB.

In the US, the most common screening test for TB is the Mantoux tuberculin skin test. The test involves injecting a small amount of TB antigens under the skin. If the individual had been exposed to the TB bacterium, there will be reaction to the antigens two days later. This reaction will appear as a red bump at the site of the injection. The size of the bump affects the diagnosis. This test diagnoses both active and latent TB.

There are a wide range of estimates for the specificity and sensitivity of the test depending upon the person's life history. This is especially true if the person is a recent immigrant from an underdeveloped country. In many regions of the world, children are vaccinated with BCG to prevent TB. The BCG vaccination produces a high rate of false positives for the skin test. For the purposes of this first example we will assume the specificity is 0.95 and the sensitivity is 0.98. The individual being tested is part of the US population that has a rate of 4% latent TB.
   a. What do sensitivity and specificity mean in this case?
   b. Why do you think medical professionals may test a healthy person for this disease?
   c. What may be some of the consequences of a false negative? A false positive?
   d. A healthy person receives a positive TB skin test. What is the probability the person actually does NOT have TB?
   e. A healthy person receives a negative TB skin test. What is the probability that the person actually has TB?

2. A clinic screens approximately 5,000 healthy individuals for TB every year using the Mantoux tuberculin skin test.
   a. On average, how many individuals will test positive at the clinic per year?
   b. On average, how many of the individuals with a positive test actually have TB bacterium?
   c. On average, how many of the individuals with a positive test do NOT have TB bacterium?
   d. How do you think a patient would react to receiving the news that he/she had a positive TB test? Do you think understanding the probabilities involved would be helpful to both the patient and doctor? Explain your answer.

3. Imagine a clinic in the developing world that screens 20,000 individuals per year. The percentage of TB infections in that country is 30%. The country does not vaccinate

against TB. This clinic uses the Mantoux tuberculin skin test. Assume the specificity and sensitivity as given above.
   a. On average, how many individuals will test positive at the clinic per year?
   b. On average, how many of the individuals with a positive test actually have TB bacterium?
   c. What is the probability that an individual with a positive test actually has TB bacterium?
   d. On average, how many of the individuals with a positive test do NOT have TB bacterium?
   e. How do you think a patient would react to receiving the news that he/she had a positive TB test? Do you think understanding the probabilities involved would be helpful to both the patient and doctor? Explain your answer.

Chapter 2 Summary

What have we learned?

In this chapter, we have explored the concept of conditional probability. We studied how the estimate of the probability of an event occurring can be influenced by knowing that another event has occurred or by having relevant information. For example, knowing a person's age and or gender affects the probability estimates for short term or long-term mortality. We studied how conditional probability can be calculated from a data table. It also can be calculated in a specific problem context such as making random selections from group that includes a mix of distinct groups. Conditional probability may also be determined from historical data such as the accuracy of test results. We also learned to use several formula that are developed around the concept of conditional probability. These include formula for calculating probabilities involving

- The intersection of two non-independent events
- The partition of a sample space
- Bayes Theorem.

We also illustrated how conditional probability can lead to counter intuitive results. In one instance we compared aggregate performance of two organizations. In another we showed that positive test results do not necessarily mean a high probability of having a disease.

Terms

Collectively exhaustive events A set of events is collectively exhaustive if at least one of them must occur (i.e., the set of events includes all possible outcomes).

Collectively exhaustive sets If sets A and B are collectively exhaustive subsets of the universal set U, then $A \cup B = U$ and $P(A \cup B) = P(A \text{ or } B) = 1$.

Compound events A compound event consists of two or more simple events.

Conditional events Two events A and B are conditional or dependent if the outcome of one event does have an effect the outcome of the other.

$$P(B \mid A) = \frac{P(A \cap B)}{P(A)} = \frac{P(A \text{ and } B)}{P(A)}$$

In other words, if we know that event A has occurred, then the sample space is reduced to the probability of event A occurring, and the denominator for $P(A \text{ and } B)$ is not one but $P(A)$.

Expected value The expected value of a (discrete) random variable is a weighted average of all possible values of the variable. The weights are the probabilities of the variable having a given value.

$$E(X) = \sum \left[x \cdot P(X = x) \right]$$

False negative A negative test result despite the presence of the object of interest to the test. The rate of false negatives is one minus the sensitivity of the test.

False positive A positive test result despite the absence of the object of interest to the test. The rate of false positives is one minus the specificity of the test.

Mutually exclusive events If events A and B are mutually exclusive, then the probability of event A or B occurring is the sum of the probabilities of events A and B.

$$P(A \cup B) = P(A \text{ or } B) = P(A) + P(B)$$

If sets A and B are mutually exclusive subsets of the universal set U, then $A \cap B = \varnothing$ and $P(A \cup B) = P(A \text{ or } B) = P(A) + P(B)$. Two sets that are mutually exclusive can also be described as disjoint sets.

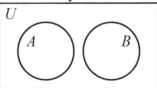

Partition of a set	If two subsets A and B are both mutually exclusive and collectively exhaustive, they are said to partition the universal set U.

```
┌─────────────┐
│U            │
│   A  │  B   │
│      │      │
└─────────────┘
```

Prevalence	The total number of cases of a disease in a given population at a specific time. The ratio of cases to population is the prevalence rate.
Prior probability	The probability assigned to a parameter or to an event in advance of any empirical evidence.
Sensitivity of a test	(True positive rate) The proportion of actual positives which a test correctly identifies as such.
Specificity of a test	(True negative rate) The proportion of actual negatives which a test correctly identifies as such.

Chapter 2 (Conditional probability) Objectives

You should be able to:

- Calculate conditional probability from a data table.

- Use conditional probability to calculate $P(A \cap B)$.

- Calculate the expected value of a random variable.

- Use the concept of a partition to calculate the total probability of an event.

- Apply Bayes Theorem to calculate the probability an individual has a disease given a positive test result.

Chapter 2 Study Guide

1. Define independence and provide an example of independent events.

2. Provide an example of events that are not independent.

3. What is conditional probability?

4. How can conditional probability be used to calculate the probability of two non-independent events occurring?

5. What is the probability an individual will live to the age of 10? What is the probability an individual who is two years old will live to the age of 10?

6. What is the formula to calculate the expected value of a random variable?

7. Write a formula that demonstrates how to use a partition to calculate the total probability of an event.

8. Define the terms false positive, false negative and prevalence.

9. What is the Bayes Theorem? Demonstrate how it can be used to explain why a positive test result for a rare disease does not necessarily mean the individual has the disease.

CHAPTER 3:

Decision Trees

Section 3.0 Decision Trees

We all make decisions in our jobs, our communities and in our personal lives that involve significant uncertainty. Some examples are:

- How much technology should be invested in a plant when product demand is uncertain?
- How much car collision insurance should be purchased when you do not know whether or not you will have an accident?
- Should a company build a new power plant in a foreign country and, if yes, in which country?
- On a personal financial level, how much should you invest in a particular stock or mutual fund?
- Should millions be invested in a new drug that has proven effective in animal tests?
- How much time should you invest in studying when the questions on the exam are uncertain and payback from more study-time is hard to predict?
- What career should you pursue when the economy and job market are uncertain?

Specific stocks and the stock market as a whole demonstrate much uncertainty from day to day and year to year. There is uncertainty in the demand for power and the stability of developing countries. The link between animal drug trials and drug effectiveness in humans is far from perfect. Individuals experience different types of random accidents.

Decision analysis is an operations research modeling tool used to select the best decision in the presence of uncertainty.

- What specific uncertainties do you face in the next day, week, or month?
- What about other members of your family or friends?
- What uncertainties will affect your planning over the next year?

The oil industry was one of the earliest users of decision analysis and continues to lead in its application. Pharmaceutical companies apply decision trees, which are an extension of probability trees, to make research and development decisions. Industrial giants such as DuPont, Xerox, and Kodak have used decision trees to plan new products and production capacity. The US Forest Service uses decision trees to plan controlled forest fires. The Decision Analysis Affinity Group (www.daag.org) is an organization that runs conferences at which corporate users of decision analysis share experiences. There is also the Decision Analysis Society which is an organization affiliated with the Institute for Operations Research and the Management Sciences (INFORMS). They maintain a homepage at https://www.informs.org/Community/DAS

The decision tree methodology involves accounting for every possible decision and random outcome. The best alternative generally maximizes the ***expected value*** of profit or minimizes the expected value of cost. Sometimes the variable optimized is not financial. The goal might be to minimize time to complete a task or maximize the number of people attending an event. Modern software such as Precision Tree, an Excel add-on, helps with the analysis and offers visual representations of the results. These enable a decision maker to explore the strengths and weaknesses of the alternatives.

As decision analysis developed, the leaders in the field recognized two critical psychological and practical issues that needed to be addressed in order to make the tool of greater practical value. First, the models required estimates of probabilities that were often not easy to obtain with a detailed analysis of data. Therefore, subject matter experts were interviewed in order to estimate the probabilities. Decision analysts, along with mathematical psychologists, became leaders in the effort to understand biases and misconceptions that individuals display when asked to make a forecast. They developed interview protocols in order to obtain expert opinion in a way that reduces the likely bias.

Second, the expected value does not capture the fact that people are often fearful of taking risks, especially large ones. This risk aversion is the foundation for all of the insurance industry and the huge market in extended warranties. Decision analysts became leaders in researching attitudes towards risk and designing a methodology called utility theory that captures this behavior.

Section 3.1 Select Prom Location

Lily Trump and Donald Tomlin were recently appointed as co-chairs for this year's senior prom. One of the big budget items is the rental of a hall. They visited four potential locations and obtained the cost and size data in Table 3.1.1. The Petty Hall was the smallest and cheapest. It would cost $1,500 and had dancing room capacity for 420 people. The largest was the Grand Hall which would cost $2,600 but could handle as many as 515 prom goers. Providence and Independence Halls were in the midsize and middle price range.

Name	Normal Capacity	Cost
Petty Hall	420	$1,500
Providence Hall	455	$1,800
Independence Hall	490	$2,200
Grand Hall	515	$2,600

Table 3.1.1: Hall capacity and cost

This year's senior class of 525 students is the largest ever. Lily and Donald worried about the possibility of every senior coming and some bringing outside guests. If that happened, even the largest hall visited would not be large enough. They also worried that if only 50% of the students attended they would be wasting a lot of money and the hall would look empty. They approached Ms. Karin Topper, the faculty advisor for the prom.

Ms. Topper had been advising prom committees for more than 15 years. She explained to the co-chairs that they should apply probability in thinking about their decision. They would have to balance the risk of not having enough room against the cost of paying for unneeded space.

Ms. Topper created an influence diagram to highlight the issue of key random events (see Figure 3.1.1).There were two main uncertainties. One uncertainty was the percent of seniors who plan to attend. The second uncertainty related to the proportion of seniors who will bring dates who are either not seniors or who attend another school. These two numbers will affect whether or not a hall will be adequate and if they will have to turn away some ticket requests.

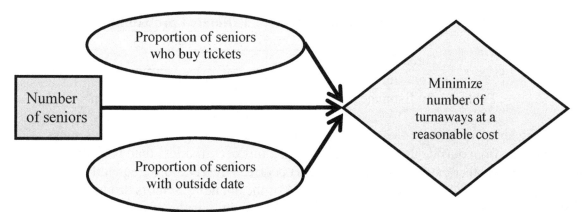

Figure 3.1.1: Influence diagram of the key random events

1. What are some reasons why a person would choose not to attend his or her senior prom?

2. At your school, what proportion of seniors would you expect not to attend prom?

3. What information could you use to estimate how many of the seniors who will attend the prom will also bring a date from outside the senior class?

Ms. Topper explained that based on her experience she could provide probability estimates for each of these two proportions. She believed that at least two-thirds of the students would attend although it could be slightly less. She thought that it was unlikely that more than four out of five would attend. She also added to the list of possibilities an intermediate rate of three-fourths of the seniors attending. In thinking about these three rates, Ms. Topper felt that with the down economy, the two-thirds and three-quarters rates were equally likely. The probability of an attendance rate as high as four out of five attending was half as likely as each of the other two rates.

Ms. Topper translated her reasoning into probabilities as follows. She let X represent the probability that the proportion attending will be two-thirds. The same value X would also apply to the three-quarters estimate. Lastly, the probability that the proportion would be four-fifths was one half of X or $0.5X$. She explained that the sum of the probabilities of the various outcomes must be 1. She used the following equation to solve for X.

$$X + X + 0.5X = 1$$
$$2.5X = 1$$
$$X = 0.4$$

Based on her experience, she estimated there was a 0.4 probability of two-thirds of the seniors attending and a similar probability that three-fourths would attend. Ms. Topper estimated there was only a 0.2 probability that as many as four out five seniors would attend. These estimates are summarized in Table 3.1.2.

Rate of seniors attending prom	Estimated probability
Two out of three	0.4
Three out of four	0.4
Four out of five	0.2

Table 3.1.2: Estimates of various rates of seniors attending prom

In thinking about 15 years of experience, she recalled that as few as 10% of the seniors brought dates from outside the senior class and that this occurred less than half the time. However, at the other extreme, the rate of 20% occurred slightly more frequently. She assigned a 0.45 probability to the event that 10% of the attending seniors brought outside guests and a 0.55 probability to the event that 20% brought outside guests.

Lily and Donald were somewhat skeptical of Ms. Topper's estimates. They asked if she had any data to back up her expert opinion. Ms. Topper told the students to speak with Ms. Sophia Numerati, the school registrar. She was known to collect all kinds of data. Ms. Numerati had in fact collected data on the proms for the last 15 years. She provided the data that appear in Table 3.1.3. Lily and Donald looked at the table. The data just looked like a bunch of numbers on a page. They needed a way to make sense of the data if they were going to use them to help make the best possible decision about which hall to rent.

Year	Class size	Seniors attendees	Percent of class	Outside guests	Percent of senior attendees with outside guests	Total attendance
1997	443	328	74%	59	18%	387
1998	476	319	67%	29	9%	348
1999	499	339	68%	64	19%	404
2000	446	290	65%	64	22%	354
2001	435	348	80%	31	9%	379
2002	486	369	76%	78	21%	447
2003	474	384	81%	77	20%	461
2004	454	341	75%	27	8%	368
2005	420	307	73%	55	18%	362
2006	494	336	68%	34	10%	370
2007	475	361	76%	40	11%	401
2008	464	302	65%	60	20%	362
2009	493	325	66%	33	10%	358
2010	478	359	75%	75	21%	434
2011	457	361	79%	43	12%	404

Table 3.1.3: Senior prom data, 1997-2011

1. Which year had the highest number of seniors attending the prom?

2. What were the highest and lowest percentages of seniors attending the prom?

3. What was the largest fluctuation in the number of outside guests from one year to the next?

4. Within the past five years, what is the range of total attendance at the prom?

5. Does there appear to be a relationship between the percentage of seniors who attend prom and the percentage of seniors who bring an outside guest?

6. Based on the data in the table, what is your best guess about how many people will attend the prom this year?

Noting their confusion, Ms. Numerati suggested that they first focus on the column that recorded the percent of seniors who attended. Perhaps they would see some patterns.

Senior Class	
Percent of Class	Grouped Average
65%	66.5%
65%	
66%	
67%	
68%	
68%	
73%	74.8%
74%	
75%	
75%	
76%	
76%	
79%	80%
80%	
81%	

Table 3.1.4: Percentage of seniors at the Prom

After reviewing the sorted data, they noticed a cluster of values in the mid-60s and another one in the mid-70s. They grouped the six lowest values and found the average for that group of six classes. The average percentage was 66.5%, which was very close to Ms. Topper's estimate of two-thirds. They grouped the values in the mid-70s and those percentages averaged out to be 74.8%, or almost three-fourths. The three highest values were all close to 80%. They used the relative frequencies for each of the categories to estimate the proportion of times a rate close to two-thirds or three-fourths was reported. These proportions could be used to estimate the likelihood or probability of that approximate rate occurring.

> **Relative Frequency and Probability:**
> There is strong two-way link between the relative frequency and probability. Analysts routinely use the relative frequency of something happening to estimate its probability. Conversely, the probability of an event happening such as rolling a seven in dice, will equal the long-term relative frequency of observing a seven.

7. How did the percentages compare to Ms. Topper's expert judgment?

They sort ordered the percentages of guests (i.e., non seniors or students from other schools) who attended as in Table 3.1.5. There once again seemed to be clusters around the 10% and 20% values mentioned by Ms. Topper.

Guests	
Percent of senior attendees with guests	Grouped Average
8%	
9%	
9%	
10%	
10%	
11%	
12%	
18%	
18%	
19%	
20%	
20%	
21%	
21%	
22%	

Table 3.1.5: Percentage of seniors who brought outside guests

8. What was the average value for each group of data?

9. What were the associated frequencies for each group? How closely did they agree with Ms. Topper's probability estimates?

> Multiply probability of two separate events: It is allowed to multiply two probabilities when the two events being modeled are independent of one another. We are assuming in this case that the percentage of seniors who choose to attend the prom does not in any way affect the percentage of seniors who will invite outside guests.

Ms. Topper suggested that the co-chairs create a probability tree to visualize the six different combinations of outcomes. They are considering only three possibilities for the proportion of seniors attending and two possibilities for the percent of outside guests they bring. The top path in Figure 3.1.2 represents the random outcome that two-thirds of the seniors attend and

10% of them bring outside dates. The probability of that happening is 0.4 × 0.45 which equals 0.18. The total number of attendees in this instance is calculated below.

$$(525)\left(\frac{2}{3}\right) + (525)\left(\frac{2}{3}\right)(0.1) = 385$$

Now consider the second path. This path involves the random sequence that two-thirds of the seniors attend and 20% of them bring guests.

10. Show that the probability of this occurring is 0.22 and the total number of attendees would be 420.

11. Calculate the probability and outcome of each of the remaining paths in Figure 3.1.2.

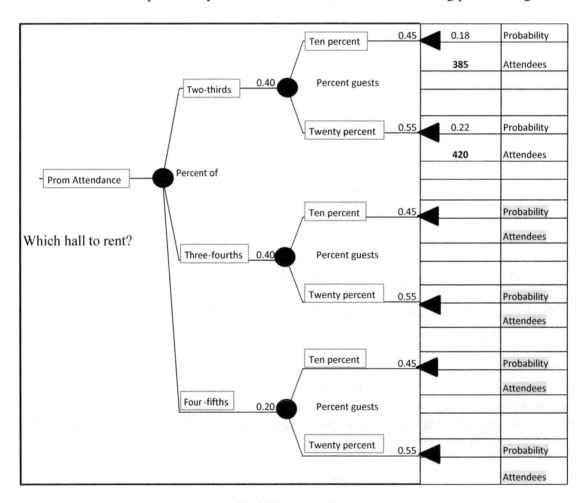

Figure 3.1.2: Probability tree of the two random events

Lily and Donald took all of the end values and their corresponding probabilities and organized them in Table 3.1.6. Before filling in the table, they sort-ordered the outcomes. The lowest value was 385 attendees and the highest was 504. In the next column they placed the associated probability. They then calculated the cumulative probabilities and recorded them in the final column. This column is interpreted as follows. There is a 0.40 probability that the number of attendees will be 420 or fewer, because 0.18 + 0.22 = 0.40. Similarly, there is a 0.89 probability that the total number of attendees will be 473 or fewer, because 0.18 + 0.22 + 0.18 + 0.09 + 0.22 = 0.89. If the estimates provided by Ms. Topper are accurate, it is certain that there will not be more than 504 attendees.

Number of Attendees	Probability	
	Relative	Cumulative
385	0.18	0.18
420	0.22	0.40
433	0.18	0.58
462	0.09	0.67
473	0.22	0.89
504	0.11	1

Table 3.1.6: Probability distribution of the number of guests

Ms. Topper suggested that Lily and Donald calculate the expected or average value of the *random variable*, the number of attendees. Any variable represents a quantity that can take on different values. The value of a random variable is subject to variations due to chance.

12. Explain why the number of people attending the prom is a random variable.

The *expected value* of a random variable is calculated by multiplying each possible outcome by its probability and adding all these terms together. The formula for expected value of a random variable with six different outcomes is

$$E(X) = \sum_{i=1}^{6} x_i \cdot P(X = x_i)$$
$$= x_1 \cdot P(X = x_1) + x_2 \cdot P(X = x_2) + x_3 \cdot P(X = x_3) + x_4 \cdot P(X = x_4)$$
$$+ x_5 \cdot P(X = x_5) + x_6 \cdot P(X = x_6)$$
$$= (385)(0.18) + (420)(0.22) + (433)(0.18) + (462)(0.09) + (473)(0.22) + (504)(0.11)$$
$$\approx 441 \text{ attendees}$$

The notation $E(X)$ is interpreted as "the expected value of X," where X is a random variable. The terms in the formula are of the form $x_i \cdot P(X = x_i)$, where $P(X = x_i)$ is the probability of the random variable having the value x_i.

After seeing this expected value, they were ready to sign a contract with Providence Hall. Its capacity was 455 people which was higher than the expected value of 441. However, Ms. Topper cautioned them to take a closer look at the cumulative probabilities in the final column of Table 3.1.6. She asked them to consider the following questions.

13. If they rent Independence Hall, what is the probability that it will be able to handle all of those who want to attend with their guests?

14. If they rent Independence Hall, what is the probability that they will have to turn away classmates who want to buy tickets to the prom?

They decided to use the information in Table 3.1.4 and add a new column to Table 3.1.1. This column would compare the capacity to the cumulative probabilities. See Table 3.1.7.

Name	Normal Capacity	Cost	Probability it will be adequate
Petty Hall	420	$1,500	0.4
Providence Hall	455	$1,800	0.58
Independence Hall	490	$2,200	0.89
Grand Hall	515	$2,600	1

Table 3.1.7: Hall options and likelihood that it is adequate

15. Which hall should Donald and Lily rent? Justify your recommendation.

Chapter 3 — Decision Trees

Section 3.2 Boss Controls Invests in Automation

Boss Controls (BC) is an automotive supplier. It manufactures integrated cup holders with temperature controls for car-makers throughout the world. Their new model is to be made available on 1,000,000 new luxury cars worldwide. Initial estimates that car buyers will select this option are as high as 50% of the time or as low as 30% of the time. Because of a general decline in the global economy, BC marketing estimates there is a slightly higher chance that demand will be 30% rather than 50%. Management assigns a 0.55 probability that only 30% of the luxury car buyers will request their temperature controlled cup holders. The only other possibility they are considering is that 50% of luxury car buyers will request this option. They assign a 0.45 probability to that possibility.

1. Why did they have to assign a 0.45 probability to the second possibility?

3.2.1 Decision

Like most companies, BC wants to maximize its profit. BC's revenue comes from selling their cup holders to car companies at a fixed price of $60 apiece. BC can control manufacturing costs by how heavily it invests in automation. An investment of $13 million in high-speed equipment would lower the cost of producing each cup holder to $12. With a more modest investment of $8 million, the cost would be $27 per cup holder. Therefore, the net profit per unit after a high or low investment in automation will be $48 or $33, respectively. This per unit profit does not yet include the investment cost.

2. What decision must be made in this situation?

3. What is the *chance event* that BC faces? (i.e., an event whose likelihood must be predicted using probability and is outside the control of BC's management)?

3.2.2 Decision Tree to Solve the BC Problem

In situations such as this, managers need to take a systematic approach to determine the best strategy for achieving their goal. We will build a *decision tree* similar to the probability tree in the last section to model the situation. The tree will show all the possible outcomes, their probabilities and their consequences.

Decision trees are made of *nodes* and *arcs*. The nodes on a decision tree represent decisions or random events. The two types of nodes, decision and random chance, are represented differently in a decision tree. Decision nodes are represented as rectangles as in Figure 3.2.1. Random chance nodes as circles as in Figure 3.2.2. There are arcs that connect these nodes. A sequence of nodes and arcs is a path that represents a possible scenario that can result from a specific decision and a random event.

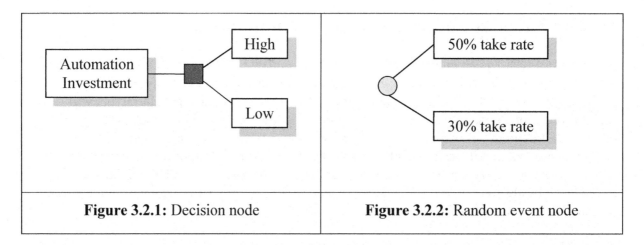

Figure 3.2.1: Decision node	**Figure 3.2.2:** Random event node

In the BC scenario, whether there will be a high or low level of investment is a *decision* that management must make. However, whether the demand for their product will be 50% or 30% is strictly a *random event*; BC's management has no control over it. The two elements, decision and random chance, can be combined into a decision tree that models the entire Boss Control scenario, as shown in Figure 3.2.3.

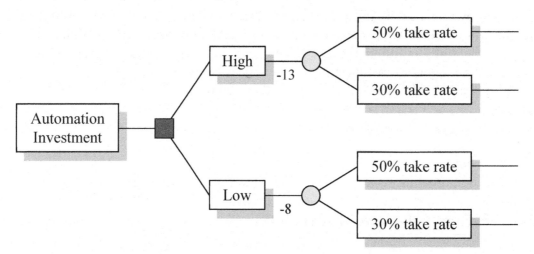

Figure 3.2.3: Decision tree for BC automation investment

This decision tree shows all the possible outcomes for every combination of decisions and random events. Now we must add the appropriate data to determine the expected profit of each path through the decision tree.

Recall that the low investment in automation was going to cost BC $8 million; the high investment was going to cost $13 million. This information can be added to the decision tree by placing the numbers along the appropriate branches, below the arcs. (The space above the arcs is reserved for another quantity, as you will see shortly). In this problem, the units on all such numbers will be millions of dollars. The numbers below are negative because they represent expenses BC must pay, not revenue earned.

Recall also that the unit profit after a low investment in automation is $33 whereas the unit profit after a high investment is $48. The "30%" and "50%" on the arcs after the chance nodes represent the proportion of car buyers who select the luxury cup holder option. This is referred to as the *take rate*. In BC's case, the estimated low take rate is 30% and the estimated high take rate is 50%.

4. What is the estimated probability that the take rate will be 30%?

5. What is the estimated probability that the take rate will be 50%?

6. What is the base of these take rates? In other words, what is the total number of potential sales?

The net revenue for each case can be calculated by multiplying the projected number of buyers times the net profit per unit. The projected net revenue with a 50% take rate and *high* investment is

Net Revenue = (1,000,000)×(0.5)×($48) = $24 million.

7. What is the projected net revenue with a 30% take rate and *high* investment?

8. What is the projected net revenue with a 50% take rate and *low* investment?

9. What is the projected net revenue with a 30% take rate and *low* investment?

In Figure 3.2.4, the earlier decision tree has been updated to include the information above. As before, the numbers beneath the arcs are in millions of dollars. They represent cost or net revenue. The numbers assigned above the arcs coming from a random event show the probability of that path.

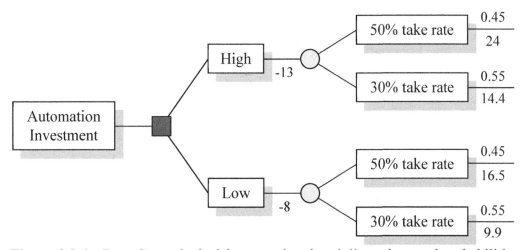

Figure 3.2.4: Boss Controls decision tree showing dollar values and probabilities

An end node, represented by a triangle, is placed at the end of each path through the decision tree. Next to the triangle, we record the net profit for that path. The net profit is determined by subtracting the investment cost from the net revenue. For example, with a high investment of $13 million and a 50% take rate, the net profit is

$24 million – $13 million = $11 million.

Figure 3.2.5 has been updated to show each of the four possible net profit values to the right of the end nodes.

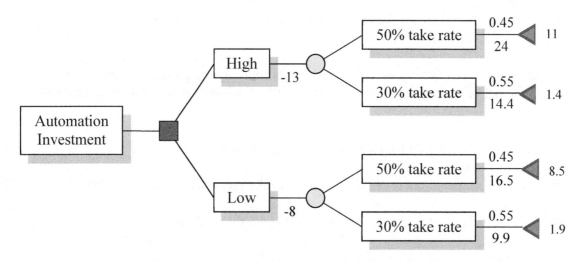

Figure 3.2.5: Boss Controls decision tree updated with values at end nodes

The next step is to determine the expected value of the net profit for each decision. We simply use the basic formula for expected value. The expected value of the net profit for a high investment is calculated as follows:

If we make the high investment decision, there is a 0.45 probability that the net profit will be $11 million. There is a 0.55 probability that the net profit will be only $1.4 million. Thus, the expected value of the net profit for the high investment decision is

Expected Net Profit = E(N) = (0.45)($11 million) + (0.55)($1.4 million) = $5.72 million

10. What is the expected value of the net profit for a low investment?

11. Is it clear from your answers whether BC should make a low or high investment in automation? Why or why not?

12. Which of the two options, high or low investment in automation, has the larger expected value for net profit? How much larger than the other option is it?

13. Suppose the vendor for the high-speed automation has just announced a $500,000 price increase from $13 million to $13.5 million. Would this affect the preferred decision? What if the increase was $1,000,000?

3.2.3 Boss Controls Sensitivity Analysis

The management at Boss Controls now wonders how sensitive the solution to their automation problem is to their probability estimates of the take rate. Recall that they have estimated the probability of a 50% take rate to be 0.45 and the probability of a 30% take rate to be 1 − 0.45 = 0.55. But, they wonder, "What if our estimate of the probability of a 50% take rate is too high? Could that change which alternative has the larger expected value of profit? If so, at what point does the larger expected value change from high investment to low investment?" In other words, they want to learn how much the probability of a 50% take rate could change and still have the high investment in automation as the preferred alternative with the larger expected value of the profit.

14. Notice that the BC managers are not wondering what would happen if their estimate of the probability of a 50% take rate is too low. Why not?

The decision to make either a high or low investment in automation rests on the expected value of the profit for each of the alternatives. Recall that those expected values are given by the calculations shown below.

Expected profit for high investment = (0.45) ($11 million) + (1 − 0.45) ($1.4 million)
= $5.72 million

Expected profit for low investment = (0.45) ($8.5 million) + (1 − 0.45) ($1.9 million)
= $4.87 million

Now, for each alternative, we can define a function to calculate the expected value of the profit based on varying probability estimates.

Let: x = the probability of a 50% take rate,
y_1 = the expected profit for high investment, and
y_2 = the expected profit for low investment.

Then, $1 - x$ = the probability of a 30% take rate,
$y_1 = (x)(11) + (1-x)(1.4)$ and
$y_2 = (x)(8.5) + (1-x)(1.9)$.

Notice that in both cases we have dropped the "$" and the "million." We can do this, because we are using the same unit of measure, millions of dollars, in each case.

Figure 3.2.6 shows the set-up and calculator graphs of these two functions on the same coordinate axes. Use a graphing calculator to graph this system of equations on the viewing window shown.

Figure 3.2.6: Expected value functions, window, and graphs for sensitivity analysis

15. The viewing window for the graphs has been set up so that x varies between 0 and 1, only. Why does that make sense for this problem?

16. What do the y-values of the functions represent? What are their units of measure?

Use the TRACE feature of your calculator to trace on the graph of y_1 until the value of x is 0.45 when rounded to two decimal places.

17. When $x \approx 0.45$, is y_1 above or below y_2? What does that mean in the context of the problem?

18. Use the calculator to find the expected value of profit for high investment if the probability of a 50% take rate is 0.4? What is the expected value of profit for low investment for this probability of a 50% take rate?

19. What are the slopes of y_1 and y_2, the expected value of profit lines? What do these slopes mean in the context of the problem?

20. Notice that the graphs of the two functions intersect. What is the significance of the point of intersection?

21. Use the calculator to find the x-value of the point of intersection, rounded to two decimal places. What does this x-value tell you?

22. For what probabilities of a 50% take rate does the high investment in automation produce the higher expected value of profit? For what probabilities does the low investment produce the higher expected value?

23. Set up an equation to determine the x-value of the intersection point.

24. Use the equation to determine the x-value of the intersection point.

Section 3.3 Green Tree Energy – Locates New Plant

Green Tree Power, Inc (GTP) is planning to expand their energy company by building a new power plant in a developing country. After much consideration, they have narrowed their possibilities to the countries of Cassedonia and Kisanthia. Each country can provide the required land and utilities for GTP to build and run their new power plant. In turn, the selected country will gain the benefits of the new energy technology that GTP can provide. Choosing the ideal location relies on several key pieces of information.

GTP estimates that the investment cost will be $50 million to build the new power plant in Cassedonia. However, there is significant uncertainty with regard to increased demand for power. As a result, the predicted total net revenue over the next five years is uncertain. Experts project that revenues could be as low as $80 million or as high as $110 million. The specific probabilistic forecast is that five-year total net revenues will be $80, $90, or $110 million with a 30%, 40%, and 30% chance, respectively. The political structure in Cassedonia is in transition. There are multiple political parties fighting for control of the country. These political parties have very different social and economic plans. Thus, the leadership of GTP believes there is a 20% chance that a Cassedonian government will take over the new power plant. If so, the government will simply repay the original investment cost of $50 million with no interest. In that case, GTP would have gained no net revenue from its investment.

If GTP builds the new power plant in Kisanthia, the investment cost is still $50 million. The population of Kisanthia is slightly lower than in Cassedonia and demand is still uncertain. GTP estimates the total net-revenue for a five-year period in Kisanthia, after the operation costs, will be $66, $80, or $90 million with a 30%, 40%, and 30% chance, respectively. The country of Kisanthia is a long established stable democracy that is committed to encouraging foreign investments. While the total forecasted revenue in Kisanthia is significantly lower than in Cassedonia, it has a major advantage. There is little chance that the Kisanthia government will take over the new power plant.

Where should GTP build their new power plant?

3.3.1 Develop the Decision Tree

Joe Riden, a senior risk analyst from GTP, has been charged with making this decision. He wants order to weigh the options carefully and objectively. Joe will create a decision tree to analyze the situation and determine the best choice for GTP.

In this situation, there is one decision to make with two possible options—GTP's new power plant can be built in Cassedonia or Kisanthia. Joe decides to develop the decision tree in stages. First he places the two alternative decisions at the first decision node. The uncertainty with regard to Kisanthia involves just the net revenue. He therefore, adds the random event, net revenue, to the Kisanthian branch of the tree. He includes all of the critical information at the appropriate places. He places a -50 on the decision branch to represent the investment cost. The net revenue random event has three branches, high, medium, and low. For each branch, Joe inputs the probability and the net revenue values. For the High revenue branch the probability is 0.3 and its projected revenue for Kisanthia is $90 million. He also inputs the corresponding numbers for Medium (0.4, $80 million) and Low (0.3, $66 million). This first stage of Joe's tree construction is presented in Figure 3.3.1.

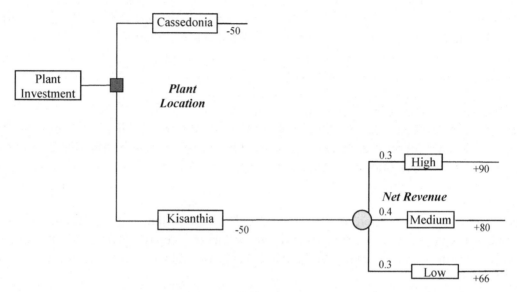

Figure 3.3.1: GTP – Kisanthian alternative

Next Joe adds the uncertainties related to building a plant in Cassedonia. Recall there is a 20% chance that the Cassedonian government will take over the power plant after it is completed. We therefore need a chance node off of Cassedonia to represent this possibility. The random event, a government takeover, has two branches: yes and no. If the government seizes control of the power plant, they will repay the $50 million dollar investment and GTP will no longer be involved with the plant. This return of $50 million is included along the Yes branch for a government takeover.

However, there is an 80% chance that the government will not take over the new power plant and GTP can move on to producing power, and thus revenue. Down this branch, there is another random event, uncertain net revenue. Joe attaches another random node with three branches for the net revenue. Again, each branch has a probability and dollar amount for net revenue: High (0.4, $110 million) Medium (0.4, $90 million) and Low (0.3, $80 million). Joe added the probability of a takeover and the net revenue uncertainty as presented in Figure 3.3.2.

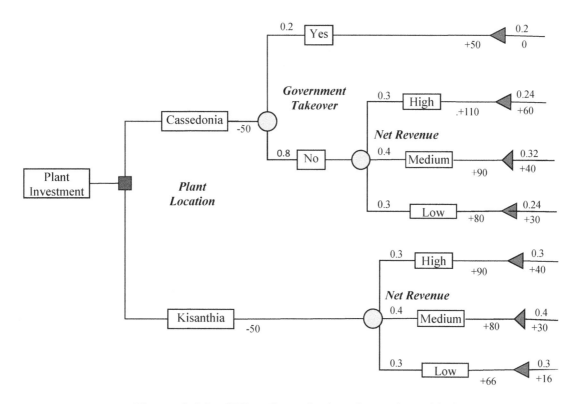

Figure 3.3.2: GTP – Cassedonian alternative added

In order to make the decision, we need to perform some calculations. At each end node, we want to note two important values: the probability of following that path from start to finish and the total profit for GTP if they follow that path. For example, assume GTP were to build in Kisanthia and net revenues turn out to be high. The net profit would be $40 million: $90 million in net revenue minus $50 million investment. The probability of that happening is just 0.3. He then adds the probabilities and net profits for the other two branches.

For Cassedonia, if the government takes over the plant, the net profit is zero. This has a probability of 0.2 of occurring. If the government does not take over the plant and the net revenue is high, the net profit is $60 million ($110-$50). The likelihood of this happening is the product of two probabilities. Mr. Riden multiplies the probability of no government takeover, 0.8, by the probability of high net revenue, 0.3. The product is 0.24. He repeats this process for the other two possible branches for net revenue for Cassedonia. Figure 3.3.3 contains all of the final dollar values and probabilities placed alongside the end nodes. The probability is placed above the net profit value.

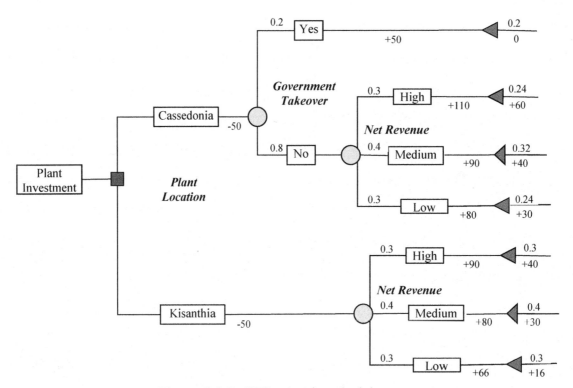

Figure 3.3.3: GTP complete decision tree

3.3.2 Compute the Expected Value

The final step in making the decision is to determine the expected value of the net profit for each alternative decision. We need the expected value of the net profit for building the power plant in Cassedonia and the expected value of the net profit for building in Kisanthia. The country with the larger expected value will be the best option for GTP.

The expected value for net profit of building the plant in Kisanthia is determined by taking the weighted sum of the different possible net profit values. There is a 30% chance of making $40 million, a 40% chance of making $30 million, and a 30% chance of making $16 million. So, the expected value of net profit is calculated as shown below.

$$E(N) = (0.3)(\$40 \text{ million}) + (0.4)(\$30 \text{ million}) + (0.3)(\$16 \text{ million}) = \$28.8 \text{ million}$$

Consider the option of building in Cassedonia. There is a 20% chance of making $0, a 24% chance of making $60 million, a 32% chance of making $40 million, and a 24% chance of making $30 million. Thus, the expected value of net profit for building the plant in Cassedonia is shown below.

$$E(N) = (0.2)(\$0) + (0.24)(\$60 \text{ million}) + (0.32)(\$40 \text{ million}) + (0.24)(\$30 \text{ million})$$
$$= \$34.4 \text{ million}$$

1. Based on the expected values, which country is preferred for the investment?

2. What is the minimum net profit for Cassedonia? How likely is that to occur?

3. What is the minimum net profit for Kisanthia? How likely is that to occur?

4. How might the issue of risk affect your preferred decision?

3.3.2 GTP Considers Insurance

Freud's of London understands the psychology of risk. It offers specialty risk insurance for large projects. Freud's is prepared to offer GTP insurance against a possible government takeover in Cassedonia. They are prepared to charge GTP a premium of $3.5 million to insure against a government takeover. If the government takes over GTP plant, Freud's will pay GTP $10 million dollars. We will analyze this insurance policy both from GTP's perspective and that of Freud's.

The GTP part of the tree is revised and presented in Figure 3.3.4. A negative $3.5 million dollars for the insurance premium is added to the cost of the Cassedonia branch. That total cost is now $53.5 million. As a result each end value for NO government takeover decreases by $3.5 million. However, the end node for the YES government takeover now has a net profit of $6.5 million dollars. They receive $50 million back from government plus a $10 million payment from the insurance company for total revenue of $60 million. The cost of investment plus the insurance premium is $53.5 million. The net profit would be $6.5 million.

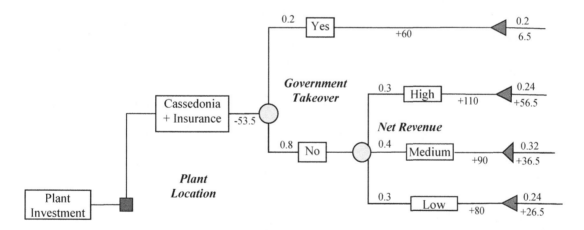

Figure 3.3.4: GTP Cassedonia with insurance

The new expected value of net profit for the Cassedonian option can be calculated. The equation shown below utilizes the abbreviation $6.5M to represent 6.5 million dollars.

$$E(N) = (0.2)(\$6.5M) + (0.24)(\$56.5M) + (0.32)(\$36.5M) + (0.24)(\$26.5M) = \$32.9M$$

The expected net profit has declined by $1.5 million. However, GTP is assured now of making at least $6.5 million from its $50 million investment in Cassedonia.

5. Would you recommend buying the insurance?

The above analysis focused on GPT's perspective. Let's look at the decision from the insurer's perspective, Freud's of London. If they offer insurance, they face only one uncertainty, a government takeover. They are not impacted by the uncertain event, net revenue. They are only insuring against a government takeover. Their decision tree is presented in Figure 3.3.5. If they sell no insurance, they gamble no money and they make no profit. If GPT purchases insurance and there is a government takeover, Freud's must pay GPT $10 million. Their net loss would be $6.5. With no takeover, their profit is the $3.5 million insurance premium they charged.

6. How much risk does Freud's face?

7. What is the expected value of the profit Freud's will earn?

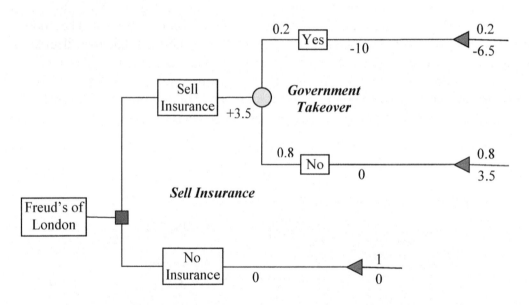

Figure 3.3.5: Freud's of London insurance decision

The early decision analysts recognized that expected value alone did not represent the way many individuals and companies deal with risk. Many of us are "risk averse." This means we would be willing to accept a reduced expected value in exchange for more certainty. This is the reason people buy insurance. They are willing to pay money to avoid risk. The cost of the insurance is always more than the expected value of the loss.

We pay hundreds of dollars to insure against a relatively unlikely catastrophic risk to our homes. We buy family medical insurance for thousands of dollars to cover the cost of possible major surgery and a long hospital stay that could cost more than a hundred thousand dollars. As individuals living one life and facing one situation, we cannot rely on long range expected value. However, insurance companies can tolerate these types of risks and live by the expected value. They can pool the risk across thousands of customers each year. Their financial performance will approximate the expected value.

Decision analysts developed a concept called utility theory to quantify this concept of risk aversion. The mathematics of utility theory is beyond the scope of the course. Instead, we will present tradeoffs between reduced expected value and more certainty and let you judge your preference. In the next example of automobile collision insurance, we will challenge you to revisit your own attitude towards risk when making smaller insurance decisions

Section 3.4 Purchase Collision Insurance

Jee Min is a high school junior at Cassidy High School in Thomasville, Michigan. He has been an excellent driver for one year. With the help of his parents, he has just purchased a 2010 Chevrolet HHT. Jee drives his car to school and to his part-time job on the weekends and after school. He is considering purchasing collision insurance for this car. It has a *Kelly Blue Book* value of $6,600. Jee gathered quotes from insurance companies through the internet. He learned that the lowest six-month premium for collision insurance with a $500 deductible is $1,700.

He is not sure that he can afford that much, so he decides to investigate the cost of collision insurance with a $1,000 deductible. The six-month premium for collision insurance with a $1,000 deductible is $1,500. The deductible represents the maximum amount of loss the owner incurs in the case of an accident. With a $500 deductible, the owner must absorb the cost of the first $500 of damages. For example, if the damages were $250, he bears the whole cost. If the damages were $6,600, he absorbs the $500 and the insurance company pays him $6,100. Jee Min is also considering the possibility of not carrying any collision insurance. In order to make the best decision, he must consider all of the consequences of each possibility.

To examine his options, Jee created a decision tree showing the three collision insurance possibilities. Each branch of the tree is labeled. Let's examine his decision tree. The tree begins with a decision node. Jee must decide whether to purchase collision insurance with a $500 deductible, a $1,000 deductible, or no collision insurance at all. The next node on each of the three branches is a probability event: a collision in the next six months. He chose a six-month time period for this event, so that it matches the six-month period covered by the premium quotes that he obtained. Next, Jee Min needed an estimate of the probability that he would be involved in a collision in the next six months.

After some Internet research, Jee Min learned that according to the National Highway Safety Administration, the probability that a male teenage driver in the U.S. will have an accident in any six-month period is 30%. He decided to use this probability estimate. Then he attached branches to the probability nodes to account for the possibilities that he will have an accident or not during a given six-month period.

1. How do you think the National Highway Safety Administration determined the probability a male teenage driver will have an accident in a six-month period?

2. Why was 70% assigned to the branches representing the event that Jee Min will not have an accident in the next 6 months?

Let's explore the structure of the tree (See Figure 3.4.1).

Chapter 3 Decision Trees

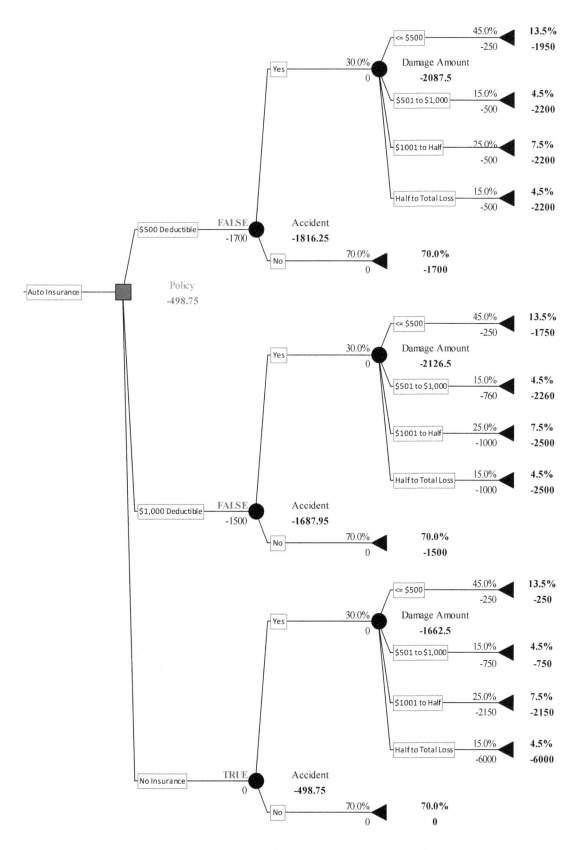

Figure 3.4.1: Collision insurance decision tree

3. There are three branches that leave the main rectangular decision node. What do these three branches represent?

4. Each of these three branches leads into a random circle node. What does that random node represent? Why are there two branches coming out of this node?

If Jee Min does have an accident, various damage amounts are possible. He realized that even if he rounded damage amounts to the nearest dollar, there would still be $6,600 possible damage amounts. Therefore, he decided to list these amounts in ranges. He also investigated the probabilities that the damage amount of an accident falls within that range. Jee Min also collected this probability data online from the National Highway Safety Administration. He organized this information in Table 3.4.1.

Damage Amount Ranges	Probability
Less than or equal to $500	45%
Greater than $500 but less than or equal to $1,000	15%
Greater than $1,000 but less than or equal to one-half the value of the car	25%
Greater than one-half the value of the car but less than or equal to $6,600, the total value of the car	15%

Table 3.4.1: Damage estimate probabilities

Based on these data, Jee Min added this information to the decision tree. He added another probability node to each branch of the tree that represents having an accident in the next six months. Then he added a branch for each of the possibilities in Table 3.4.1 to each new node. Finally, he labeled each new branch with the range of an amount of damage. He also assigned the corresponding probability of occurrence. Note the number of sequences of branches in Jee Min's tree. The probability of that sequence of random events is listed at the end of each sequence of branches.

5. Each branch labeled YES contains another random node. What does this represent? Why are there four branches leaving this second node?

Let's explore the probabilities on the tree. There is a probability assigned to the end node of each sequence of branches. For example, the top end node has a 13.5% assigned to it.

6. How was this end probability determined?

7. Which end nodes have the highest probabilities and why?

8. The probabilities for the top five branches in the picture sum to 1. Why is this the case?

Let's discuss the dollar amounts of the tree and the end node values. Jee Min realized that for each complete decision branch of his tree, he must assign a cost to each branch. However, in order to do these calculations, he realizes that he needs *individual* damage amounts and not

Chapter 3 | Decision Trees

ranges of damage amounts. Jee Min decides to use a single number within each range to represent the damage amount for that branch. He listed those in the following table.

Damage Amount Ranges	Representative Amount
Less than or equal to $500	$250
Greater than $500 but less than or equal to $1,000	$750
Greater than $1,000 but less than or equal to one-half the value of the car ($3,300)	$2,150
Greater than one-half the value of the car but less than or equal to the total value of the car	$6,600

Table 3.4.2: Damage estimate – representative amounts

For the three lowest damage ranges, Jee Min decided to use the midpoint of the range: $250 for accidents having damage less than or equal to $500, $750 for accidents having damage greater than $500 and less than or equal to $1,000, and $2,150 for accidents having damage greater than $1,000 and less than or equal to $3,300, one-half the $6,600 value of his car. For the last range, damage greater than one-half value and less than or equal to the full value of his car, he learned that insurance companies almost always "total" the car when the damage falls within this range. Therefore, he decided to use $6,600 for accidents in that range.

All four of these dollar amounts are entered onto the appropriate branches of the tree for the no collision insurance decision. Now let's look at the section of the tree with the decision, $1,000 deductible, the random event "yes", and the random event "damage". The dollar amounts for these four branches do not match the values in the table.

9. Which two branches match the table and which two do not? Why?

10. For the section of the tree with a $500 deductible, there are three branches with a $500 value. Why is this the case?

Consider only the portion of the entire decision tree that represents the decision to purchase $500 deductible collision insurance. Recall that the premium for this deductible amount is $1,700. For the top-most branch, the cost is $250. Thus, the end node value is the sum of these two costs, $1,950. The second branch, has a $500 cost. The end node value is therefore $2,200.

11. Why are the end node values for the third and fourth branch the same $2,200 as for the second branch?

In a similar way, Jee Min adds the cost of the premium with his liability for damages to determine the end node value for each sequence of branches.

12. What is the expected cost to Jee Min of the decision to purchase $500 deductible collision insurance?

13. Examine each of the remaining decision options and the total cost. In a similar way, determine the expected cost to Jee Min of making that decision. Enter each of the expected costs in the table below.

Decision	Expected Cost
$500 Deductible	
$1,000 Deductible	
No Insurance	

Table 3.4.3: Expected costs

14. Based on this analysis, which option has the smallest expected cost?

15. Should Jee Min base his decision only on this analysis? Explain why or why not.

The basis for the insurance industry recognizes the concern people have with incurring huge costs associated with relatively infrequent events. Every reduction in premium fees can be directly subtracted from the total expected value to determine the impact.

16. Given your own attitude towards risk, would you be willing to pay more than the expected value calculated for the no insurance options to reduce your risk? If so, how much would you be willing to pay to have a policy with a $1,000 deductible?

3.4.1 A Revised Estimate of the Probability of Having an Accident

Over the past six years, Michigan and other states in the U.S. have instituted graduated driver training programs for new teenage drivers. In fact, Jee Min participated in such a program. The establishment of these programs has resulted in safer teenage drivers.

What if the National Highway Safety Administration has now determined the probability that a teenager who graduated a driver trainer program will have an accident in any six-month period is 22%.

17. Will this new estimate of the probability of having an accident affect the expected cost for each decision?

18. Using this new estimate, what is the probability of *not* having an accident in the six-month period?

The insurance companies are slowly responding to this reduced rate of accidents. They are considering significant reductions in the premiums charged.

19. If the premiums were reduced by 25%, what would be the expected values for each of the insurance policies?

20. Recalculate the expected cost for no insurance option. Should Jee Min change his decision? Explain.

Chapter 3: (Decision Trees) Homework

1. The probability of selling a dress in a store is 25% each week.
   a. Construct a probability tree to determine all of the possible outcomes over a three-week period.
   b. What is the probability that the dress will not be sold at the end of the third week?

2. A contestant on a TV show must pass four stages to win a big prize. The probabilities of winning in stage 1, 2, 3, and 4 are 0.8, 0.6, 0.4, and 0.3 respectively. The contestant wants to know the probability of winning the big prize.
   a. Construct a probability tree to determine the possible outcomes of the game.
   b. What is the probability that he wins the big prize?
   c. What is the probability that he makes it to stage 4 but does not win the big prize?

3. A TV cable company has a technical support department to solve customers' problems by phone. In this department the staff members are categorized into four levels based on their ability to solve customer problems. The company first assigns a problem to Level 1; if they cannot solve it, someone at Level 2 is assigned. This process is repeated until it finally reaches the most experienced staff for one last attempt at solving the problem. The probability that a staff person is able to solve the problem at each Level is 0.50, 0.75, 0.85, and 0.95 consecutively.
   a. List all of the possible outcomes.
   b. Construct a probability tree to determine the probability of each outcome.
   c. What is the probability that a customer's problem is unsolved?
   d. What is the probability that the problem is solved by someone at Level 3?
   e. Why is this probability less than the probability that a Level 1 individual solves the problem?

4. A manager at Wayne State football games must decide 10 days in advance which product to order for the stadium vendors to sell. Each product will have the university logo. The three options are sunglasses, umbrellas, and ponchos. He will stock only one of the items. Sales and the resultant profit will depend upon the weather on the day of the game. The long-range weather forecast is 35% chance of rain, 25% chance of overcast skies, and 40% chance of sunshine. Table 1 contains the manager's estimates of the profits that will result from each decision and each weather condition.

Decision	Profit ($)		
	Weather Condition		
	Rain (0.35)	Overcast (0.25)	Sunshine (0.40)
Sunglass	-600	-300	1,600
Umbrella	2,100	0	-800
Poncho	1,800	500	-600

Table 1: Wayne State sales of items with logo

a. What is the best decision for each weather condition?
b. Draw the associated decision tree needed to make the best decision.
c. What decision should be made if he desires to maximize the expected value?

5. The owner of a restaurant is considering two ways to expand operations: open a drive-thru window or serve breakfast. There are increased annual costs which each option and a one-time cost associated with the drive-thru. Labor and marketing costs are annual costs that the restaurant has to pay each year. They include hiring new staff and placing more ads in local media. Redesigning the restaurant is a one-time cost that is paid at the beginning and does not repeat each year. The details are provided in the following table.

Decision	Costs ($)		
	Annual		One Time
	Labor	Marketing	Restaurant Redesign
Drive-thru window	28,000	10,000	20,000
Breakfast	38,500	5,000	-

Table 2: Restaurant expands operations - costs

The forecasted increase in income resulting from these proposed expansions depends on whether a competitor opens a restaurant down the street or not. Based on the restaurants evaluation, the manager believes that the competitor won't open a new restaurant with 60 percent probability. Based on the competitor action, the restaurant's profit will be different for each decision. The following table provides an estimation of the increase of income based on the competitor's action.

Decision	Revenue ($)	
	Competitor	
	Open (0.4)	Not Open (0.6)
Drive-thru window	110,000	130,000
Breakfast	80,000	120,000

Table 3: Restaurant expands operations - revenue

The owner of the restaurant is focused just on next year. He therefore decided to consider the one-time cost for the redesign the same as all of the labor and marketing costs that are ongoing.
a. Calculate the profit of each decision when considering the competitor's action.
b. What is the best alternative if no competitor opens nearby? What is the best alternative if a competitor opens nearby?
c. Draw the associated decision tree.
d. What decision should the company follow?
e. Let p represent this probability that the competitor will open a restaurant down the street. Write an equation to calculate the expected value for each decision as a function of p.
f. For what value of p are these two expected values equal?

g. Graph the equations of the expected values to determine their intersection point. What does this intersection point represent?

h. Recall that the owner treated the design change and marketing cost the same as operating costs. Would the decision change if he considered only 50% of these costs this year (design and marketing)?

6. A company is about to launch its new fast food for sale in supermarkets throughout Arkansas. The research department is convinced that a special type of chicken wings will be a great success. The marketing department wants to launch an intensive advertising campaign. The advertising campaign will cost $1,000,000 and if successful will produce $4,800,000 profit. If the campaign is unsuccessful (25% chance), the profit is estimated at only $1,800,000. If no advertising is used, the revenue is estimated at $3,500,000 with probability 0.6 if customers are receptive and $1,500,000 with probability 0.4 if they are not.

   a. Draw the associated decision tree.
   b. What course of action should the company follow in launching the new product if they want to maximize the expected value?
   c. Write an equation to calculate the expected value for each decision as a function of the probability that the major advertising campaign will be effective? (Hint: one of the alternatives is unaffected by this probability.)
   d. Graph the equations of the expected values to determine their intersection point. What does this intersection point represent?

7. A discount clothing store uses an interesting strategy to attract customers to return each week to shop. They tell the customers that every 7 days they reduce the most recent price of an unsold dress by an amount equal to 25% of the original price. On each dress there is a label of its original price and the date it was hung on the rack. Thus customers know that a $40 dress placed out on Nov. 7th will be priced only $30 on Nov. 14th if it is not sold before then. It will be reduced by another $10 on Nov. 21st if it is still unsold. After three weeks, any unsold dress is sent to a local charity. Each week, there is a .60 probability that the dress will be sold.

   a. Nancy Drew saw a dress she really liked and knows she can get the almost identical dress for $50 online. The current store price is $40. Construct a decision tree to determine whether or not she should buy the dress now or gamble and wait a week. She would buy it next week if it remains unsold. (If when she comes back next week, Nancy finds the dress has been sold, she will buy it online.)
   b. Just before finalizing her decision, she found another place online that sells the same dress for $45. Why might a lower price online affect her purchase decision in this store? Should she buy the dress now or gamble and buy it in the second week if available?

c. She just saw a more expensive dress for sale at $80. These more expensive dresses have only a 40% chance of being sold each week. Again, they tell the customers that every 7 days they reduce the most recent price of an unsold dress by an amount equal to 25% of the original price. Assume she would be able to buy a similar dress for $90 online. Construct a full tree for 3 weeks and specify the optimal decision.

8. A contestant on a TV show has to decide whether to stop or try to answer another question. The contestant is first asked a question about US Geography. If the contestant answers correctly, she earns $700. Historically, three out of four contestants answer the first question correctly. If answered incorrectly, the game is over. If answered correctly, the contestant can leave with $700 or go on and answer a question about US presidents. If answered correctly, the contestant wins an additional $1000. If the answer is incorrect, the contestant loses all previous earnings and is sent home. Historically, two out of three contestants answer this question correctly. The third question is about rock 'n' roll music. This question is worth $1500, and the same rules apply. The chance of answering this question correctly is 50-50.
   a. Draw a decision tree that can be used to determine how to maximize a contestant's expected earnings. What is the best decision and what are the expected earnings in this case?
   b. Some contestants may feel more or less knowledgeable about the third question category. Let p represent the probability that a contestant will answer the third question correctly. Write an equation to calculate the expected value for attempting to answer the third question in terms of p.
   c. Based on the previous question, what is the cutoff value of p such that a contestant should attempt the third question?
   d. The TV show is considering changing the reward for answering the 3rd question correctly. Let m represent the amount of money a contestant will earn for answering the third question correctly. Write an equation to calculate the expected value for the last decision as a function of m.
   e. Graph the equations of the expected values to determine the intersection point for the last stage. What does this intersection point represent?

9. SSS Company, a software company, is considering submission of a bid for a state government contract to install their software on 30,000 computers. The government would use their software to oversee the management of tens of thousands of large and small contracts the government signs every year. There is only one other potential bidder for this contract, Complexo Computers, Inc. Complexo has a long record and reputation with this kind of contract. As a result of its lesser experience, to win the bid SSS's bid must be at least $5 less per computer installation than Complexo's. Complexo Computers is certain to bid and is generally more expensive than SSS. SSS management believes that it is equally likely that Complexo will bid $100, $90, or $80 per computer installation.
   a. What are the possible bids that SSS should consider?

SSS's bidding decision is complicated by the fact that it is currently working on a new process to install software remotely through the internet. If this process works as hoped, then it may substantially lower the cost of installations. However, there is some chance that the new process will actually be more expensive than the current installation process. Unfortunately, SSS will not be able to determine the cost of the new process without actually using it to install the software. The higher SSS bids the more money it makes if it wins the contract. However, the higher the bid, the less likely it is to win the contract. If SSS decides to bid, it will cost $20,000 to prepare all of the relevant documents required to submit the bid. SSS will incur this expense regardless of whether it wins or loses the bidding competition. With the proposed new installation process, there is a 0.25 probability that the cost will be $50 per computer and a 0.50 probability that the cost will be $75 per computer. Unfortunately, there is also a 0.25 probability that the cost will be $85 per computer.

    b. Construct a decision tree to model this situation.
    c. Based on your decision tree, do you recommend that SSS Company submit a bid. If so, what should they bid per installation?
    d. Under the optimal policy, what is the probability they will win the contract?
    e. What is the overall expected value if they bid on the contract?
    f. If they win the contract, what is their expected value of profit? (Hint: This is a conditional decision analysis.)

10. A group of high school students has decided to start a summer business. They are thinking about designing and coloring T-shirts and selling them to clothing stores in their community. For mass production of colored T-shirts, they need special equipment which they can buy or rent. After negotiating with a company about equipment, they figure out that they have three options to start their business:

- They can buy all of the equipment and do the design and printing themselves. In this case they have to pay for the equipment but they can recover part of the money at the end of summer by reselling the equipment. The cost of buying the equipment is $8,100. They can resell it at 50% of the original price. The cost of printing will be $1 per T-shirt.

- The second option is renting the equipment and returning it at the end of summer. The renting cost is $1,500 for the whole summer with a variable cost of $1.50 per print.

- The third option is outsourcing the printing. In this case they do the designs themselves but send them to a company for printing. The company charges them $2 per T-shirt.

In each option the unprinted T shirt costs them one dollar.

The fact that the market demand for colored T-shirts is not certain makes the decision making difficult. After doing some market evaluation, they summarized their expectation in following table.

Demand (Number of T-shirts)	Probability
2,000	15%
5,000	50%
8,000	35%

Table 4: Probabilistic demand for T-shirts

    a. If they can sell each T-shirt for $5, construct a decision tree to help make the decision.
    b. What is the best option if the demand is 2,000 T-shirts?
    c. What is the best option if the demand is 5,000 T-shirts?
    d. What is the best option if the demand is 8,000 T-shirts?
    e. Which option is the best for them? What is the expected profit if that decision is made?

11. A software company released a beta version of a software package. It expects a large number of requests from the users for fixing potential bugs in the software. These include crashing, lock up, and incompatibility errors. The company has established a help desk to handle telephone requests. The company trained two groups of software specialists to support the software. Group 1 has just been hired and trained; meanwhile specialists in Group 2 are senior technicians very capable of solving the problems. The senior specialists solve the problems with 100 percent certainty, but their salaries are much higher than other specialists.

The payment system of the company for the specialists is problem based. The company pays them based on the number of the problems that they attempt to solve. Group 1 salaries are $20 per problem and Group 2 salaries are $35 per problem. The software company always has a dilemma as to which specialist to assign in order to minimize the cost of the support. For example, assume they assign a problem to a Group 1 specialist. If he is not able to solve the problem, they reassign it to a senior specialist. In this case both specialists are paid. This costs the company $55 per problem. To address this issue, they developed an automatic system to predict the chance of solving a problem by Group 1 based on previous cases.

    a. A crashing problem was just received, and the prediction software forecasts a 70 percent chance of success for a Group 1 specialist. Draw a decision tree for this problem.
    b. What kind of specialist should be assigned to the problem first?
    c. Another problem, compatibility error, was received and the prediction software forecasts a 50 percent chance of success for a Group 1 specialist. Draw a decision tree for this decision.
    d. Based on the decision tree, what kind of specialist should be assigned to the problem?
    e. Let p represent the probability that the Group 1 specialist will be able to solve the problem. Write an equation to calculate the expected value of the cost for each decision as a function of p.
    f. Graph the equations of the expected values to determine their intersection point.

g. They want to know what should be the cutoff value of the probability to assign directly to an expert instead of assigning the task to a Group 1 specialist. Use both a graphical representation of part e) and an algebraic representation of part f) to find that probability.
h. In the previous question, what was the role of the two salaries in determining the breakeven value of p? Assume that the salary of a Group 1 specialist is x and the salary of Group 2 specialist is y. Write an equation to calculate the expected value of the cost for each decision as a function of p, x and y.
i. Find the value of p as a function of x and y that leads to the same expected value of the cost regardless of whether the problem is first assigned to a Group 1 or Group 2 specialist.

12. (Continuing the previous problem) After finishing the first phase, management figured out that they need another group of specialists that are more knowledgeable than Group 1 but not necessarily experts. They are to be paid at $28 because of their higher success rates than Group 1. This group is called Group 1.5.
    a. A problem was just received and the prediction software forecasts 70 percent chance of success for a Group 1 specialist. There is an 85 percent chance of success for a Group 1.5 specialist. Group 2 experts can solve the problem for sure. Draw a decision tree for this problem. (Assume that if a Group 1.5 specialist fails to fix the problem, the company will then assign it to Group 2.)
    b. Based on the decision tree, what kind of specialist should be assigned to the problem first?
    c. What is the optimal decision if the probability of success for Group 1 would be 60% and still 85% for Group 1.5?

13. An automotive part has to go through two different processes using metal lathes to be shaped properly. Each process has a cost associated with the type of lathe. Each step in processing has a risk of ruining the part and turning it into scrap. For example, when Lathe 1 processes a part, the cost is $100 and the risk of being scrapped is 10%. Each part that successfully processed by both lathes is sold for $450. The net profit is equal to the number of parts sold minus the cost of processing all parts. The cost of processing includes both finished and scrapped parts. The following table shows the cost and risk of each lathe.

	Cost	Risk of scrap
Lathe 1	$100	10%
Lathe 2	$150	20%

Table 5: Two lathe costs and scrap

There is no recycling value if the part is ruined; scrapped parts are worthless.
    a. What is the probability that a part will end up being scrapped? Does it make a difference as to the order of the processes?
    b. The processes can be done in either order. Draw a decision tree to determine the optimal sequence of processes which maximizes the expected value of the net profit per part.
    c. Which process should be done first?

d. If the cost of processing by Lathe 2 changed, for what cost would the optimal strategy change? What is this percentage change?

14. Continuing the previous problem, suppose a different part must undergo three processes to be done by three different lathes. Parts that are made with these 3 processes are sold for $800 each. Each lathe has an associated cost and risk of ruin as follows:

	Cost	Risk
Lathe 1	$100	10%
Lathe2	$150	20%
Lathe3	$200	25%

Table 6: Three lathe costs and scrap

a. How many different sequences need to be considered?
b. What is the probability that a part is ruined?
c. Draw a decision tree to determine the optimal sequence of processes which maximizes the expected value of the net profit per part.
d. In which order should the processes be done?
e. Explain how you can use a pair-wise comparison among processes to find the optimal sequence?
f. If there were four processes, how many pair-wise comparisons would need to be made to find the optimal sequence?

Chapter 3 Summary

What have we learned?

We have learned that decision-making often involves uncertainty. It may be easy to choose between two events such as taking a job that pays $8 per hour or a job that pays $10 per hour. However, it becomes more difficult to choose if the benefits are less certain. What if the second job involved a lower base pay but you could earn tips. Imagine your base pay was $6 per hour but tips ranged from $0 to $4 per hour. Your decision would need to be based on how likely those amounts were. In order to choose between multiple options, we need to be able to identify not only the benefit of different possibilities but also the likelihood of the different possibilities occurring. A decision tree can help combine these together to find the expected values of different options. The option we should choose is the one that has the highest expected value. It is interesting to note that often the expected value is not actually a possible value.

Using probability decision trees is a multiple step process:
1. Create a decision tree including decision nodes, chance nodes and end nodes.
2. Identify the potential cost or benefit for each branch off of a node.
3. Calculate the cost or benefit for each complete path.
4. Identify the probability for each branch off of a chance node.
5. Find expected value for each chance node.
6. Compare the expected values for each branch of the decision node.
7. Select the decision node branch with the best expected value.

Terms

Arcs or branches	The connections between nodes on a decision tree showing alternative decisions or outcomes of random events.
Chance event	An event whose likelihood must be predicted using probability and is outside anyone's control.
Compound event	An event that is the combination of two or more simpler events.
Decision tree	A diagram similar to a probability tree used to model decision alternatives in which the profits or costs associated with the decision alternatives are affected by one or more random events.
Expected value	The weighted average found by multiplying the possible results of a random variable by the probability of the random variable having that value.
Fundamental principle of counting	If there are m possible outcomes for one event and n possible outcomes for a second event then there are $m \times n$ possible outcomes when both events occur.
Independent	Events A and B are independent if the outcome of event A does not affect the outcome of event B, and the outcome of event B does not affect the outcome of event A.
Multiplication rule	If A and B are independent events then the probability of A and B both happening equals the probability of A times the probability of B.
Node	The boxes (representing decisions), circles (representing random chance), and triangles (representing the end of a path) on the decision tree.
Probability tree	A diagram used to show all the possible outcomes for a combination of two or more independent events.
Random variable	A variable whose numeric value depends on random events.
Risk aversion	The desire to avoid risks that can affect either costs or profits.
Take rate	The proportion of customers that select a particular option.

Chapter 3 (Decision Tree) Objectives

You should be able to:

- Create a probability tree to calculate the probability of different sequences of events.

- Create a decision tree including decision nodes, chance nodes and end nodes.

- Identify the potential cost or benefit for each branch off of a decision node.

- Identify the probability for each branch off of a chance node.

- Use the fact that probabilities must add up to one.

- Find the value for each sequence of branches of the decision tree.

- Find the expected value for each decision option.

Chapter 3 Study Guide

1. What is a decision tree and what is it used for?

2. What are nodes on a decision tree? What shape is used for each type of node?

3. How do you find the end node value for a path in a decision tree? Where do you put that value on the decision tree?

4. Once you find the value for each branch of a decision tree, how do you find the expected value for each alternative of the decision nodes?

5. What must be true about the sum of the probabilities of the branches after a chance node?

6. What does risk aversion mean and how does it affect the results of decision tree analysis?

7. Give an example of risk aversion.

CHAPTER 4:

Binomial and Geometric Distributions

| Chapter 4 | Binomial and Geometric Distributions |

Section 4.0 Random Number of Successes and Trials: Binomial and Geometric Distributions

Chapter 1 was designed to explore randomness in a number of decision contexts. We modeled various examples using random number simulations. Then we analyzed the output of the simulated data to determine the observed relative frequency of critical outcomes. We also applied the multiplication rule to estimate the probability of a specific outcome. In each of the examples, we started with a specific probability, p, of something happening. In Koala Foods, p referred to the performance standard that there was 0.5 probability the phone would be answered at 8:02 a.m. For the high school newspaper, p referred to the 0.95 probability that a writer would submit an article on time. With regard to BT Auto, the p value of interest was the 0.1 likelihood that, on a particular day, one of the workers would be out sick due to the flu. In addition, we assumed that each repetition or individual was independent of every other and had the same p value. Thus, whether or not the phone was answered on Monday at 8:02 a.m., did not affect the probability it was answered on Tuesday at 8:02 a.m. Whether writer X met or did not meet the deadline did not influence any other writer on the team. Similarly, worker A's absence or presence on a specific day did not change the probability for each and every one of the other workers on that day or any other.

The last common element of each problem was a value n, the number of times something was repeated. You flipped a coin 48 times. Phone calls were made to Koala Foods three days in a row. There were 10 writers, each of whom was to submit an article. In the case of BT Auto there were two sets of repetitions presented in rows and columns. Each row contained the simulated data for each of the 12 workers expected to show up for work that day. For each row, n is 12. Each column had 50 repetitions for an individual worker. This corresponded to the days the worker was absent or present at work. For each column n is 50.

In summary, there were four common elements to the random event being analyzed. In each case:
1. There were only two possible outcomes.
2. The p value, the probability of success, was the same for each repetition.
3. The outcome of each repetition was assumed to be independent of every other repetition.
4. There were n identical repetitions of the random event.

4.0.1 The Number of Successes - Binomial Distribution

These four basic elements and assumptions are enough to enable us to build a probability model, a formula that can be used to calculate probabilities. This formula can complement and eventually replace the detailed simulations and associated data analysis presented in the previous chapter. The model we develop in this chapter is called the ***binomial distribution***. It has two parameters, values that need to be specified in order to apply the distribution. These parameters are n and p.

Let: n = the number of times the random event is repeated
      p = the probability that a specific outcome, called a success, occurs.

Then: $(1-p)$ is the probability of the specific outcome *not* occurring.

Mathematicians and statisticians use standard words. They call the outcome of interest a ***success*** and its complement a ***failure***. The terms success and failure are not meant to have either positive or negative connotations. If we are interested in studying writers who do not meet deadlines, the event of interest is "not meeting a deadline" and would be labeled a success in our example. However, if we are interested in counting the number of writers who do meet deadlines, the event of interest "meeting a deadline" would be labeled a success.

4.0.2 Random Variables

The variable X is called a ***random variable***. Calling X a ***variable*** means that it can take on different values. Calling X a ***random variable*** means that the different values of X occur at random.

Let X = the number of successes.

When we repeat the random event n times, X, *the number of successes,* has a range from zero to n. For example if we call Koala three days in a row, the number of times someone answered the phone is a number between zero and three. Similarly, the number of writers who meet the deadline for a specific edition of the paper can range from zero to 10 writers. In this chapter, the teacher in charge of the school paper is considering publishing the paper even if only eight writers have submitted their work. Thus he will be interested in knowing.

$P(X \geq 8)$

What is the probability that eight or more writers have met the deadline? The binomial distribution has a formula that can be easily input into your calculator to answer this question. Similarly the manager of BT Auto is considering scheduling spare workers. He wants to know the probability that two workers will be absent on a given day. If X represents the number of workers absent, the manager wants to calculate $P(X = 2)$.

4.0.3 The Number of Attempts until 1st Success – Geometric Distribution

In some situations we are interested in how long it will take to achieve a first success. This is also a random variable. Imagine you are in charge research and development for a drug company. Assume that each time your team studies a possible drug compound there is only a one-in-500 chance that it will lead to a drug that can be sold. As a manager, you are interested in the random variable, the number of compounds tested before the team finds a saleable drug.

Imagine a different context. You are in charge of planning space shuttle flights. Your engineers have estimated that there is a one-in-80 chance of a catastrophic failure. What is the probability that the first failure will occur on or before the 50th shuttle flight?

Now let's return to the examples from the previous chapter, Mr. Smith of Koala Foods is interested in the first day a call at 8:10 a.m. is not answered.

Let R = the number of times an event is repeated, until we observe a specific outcome for the first time.

The smallest possible value of R is just one. You cannot have the first unanswered phone call without calling the first time. The largest possible value of R is infinity. Koala Foods may never miss a phone call at 8:10 a.m.

The formula used to calculate probabilities for this type of problem is called the ***geometric distribution***. For example, the geometric distribution can be used to calculate the probability that the first day a call at 8:10 a.m. is not answered is on the fifth day, $P(R = 5)$. The geometric distribution is closely related to the concept of a geometric series that you studied in earlier courses. In the geometric distribution there is just one parameter, p, the probability of the outcome of interest.

4.0.4 Chapter Examples

In the previous chapter we explored the likelihood of no one answering the Koala Foods customer service line for several consecutive days. In this chapter we explore a wider range of possible outcomes. We introduce the binomial distribution as a standard for modeling situations like these. We also develop a second example based on *The Lancer* newspaper scenario. This scenario has two added features. We demonstrate how to use a cumulative distribution of the binomial distribution to answer questions. We also apply the concept of ***conditional probability***.

The second set of examples develops the geometric distribution. We use two contexts. The first explores the probability distribution of the first day on which a call to the Koala Foods customer service line is answered at 8:02 am. The second example explores the time between space shuttle catastrophes.

4.0.5 Probability Distributions and Relative Frequency

There is a direct link between the simulations we carried out in the previous chapter and probability distributions. Recall that each of your simulated data sets was often very different from that of your neighbor. When you pooled the classroom data, you had a much larger sample. In general, as you increase the number of times you simulate a random event, the frequency of specific outcomes more and more closely approaches the probability values that are calculated by the binomial distribution formula. Thus, these formulas are short cuts to estimating probabilities. They replace the need to repeat the simulations over and over again in order to obtain reliable estimates of the relative frequency of the outcome being studied. At the close of the chapter we will offer opportunities to experience a large number of repeated repetitions of the same random experiment. This will be used to demonstrate how the relative frequency approaches the theoretical probabilities calculated by the probability distribution formula.

Section 4.1 Customer Service at Koala Foods – Answered Calls

In an earlier chapter, Mr. Smith decided the data presented by AGB Company was inconclusive. Even if the 8:02 a.m. standard of 50% of calls answered was generally met, there was a one-in-eight chance that during a three-day period no one answered the phone. He decided this was not enough evidence to take corrective action. Before seeking more data, Mr. Smith decided he wanted to better understand the likelihood of different outcomes from calling the help line three days in a row. There are four possibilities as to the number of times the call was answered: 0, 1, 2, and 3. He wondered, what is the probability of each of these outcomes assuming that the 50% standard is generally achieved over the course of a year?

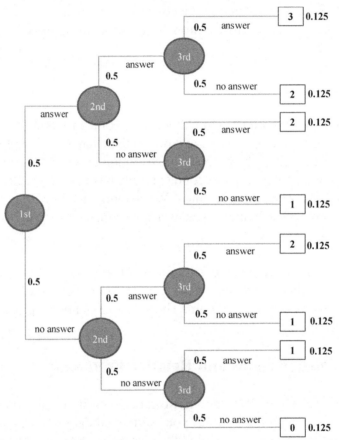

Figure 4.1.1: Probability tree – Call three days in a row at 8:02 a.m.

Mr. Smith constructed a probability tree to lay out the sequence of events that might occur when calling three days in a row. Each day there are two possible outcomes; the call is answered or it is not answered. The standard set for answering calls at 8:02 a.m. was 50%, so these two possibilities are equally likely. To develop the tree, he multiplied the number of outcomes each day for three days and determined there was a total of eight possible sequences.

This tree is depicted in Figure 4.1.1. At the end of each path, he recorded the probability of that path or sequence of events. He also recorded in a box the number of calls that were answered on that path. The top path in the figure corresponds to the call being answered every day. The probability of that happening is 0.5 × 0.5 × 0.5 or 0.125. In this path, a total of three calls were answered. Similarly, at the bottom of the figure is the sequence of days with no answer each day. It too has a probability of occurrence of 0.125 with zero calls answered. In fact, the probability of each and every sequence of three daily calls has a probability 0.125 because of the equal probability of an answer and no answer.

In thinking about the possible outcomes, Mr. Smith realized he was more interested in the number of times the phone was answered at 8:02 a.m. than in the specific sequence of what happened each day. The number of times the phone was answered is a random variable. To help with the analysis, he recorded at the end of each sequence the number of calls that were answered. He first analyzed the probability of the phone being answered exactly once. He identified the three distinct paths listed below.

Path	Day			Path probability
	1	2	3	
1	answered	not answered	not answered	0.125
2	not answered	answered	not answered	0.125
3	not answered	not answered	answered	0.125

Table 4.1.1: One call answered out of three – Mutually exclusive paths

Each of these paths has a probability of 0.125. These paths are considered mutually exclusive events even though pieces of the path overlap. They are mutually exclusive in that if one 3-day sequence occurred the other 3-day sequence did not occur. Because they are mutually exclusive, he can add probabilities together to determine the probability a call was answered on exactly 1 day.

$$P(X=1) = 0.125 + 0.125 + 0.125$$
$$= 3(0.125)$$
$$= 0.375$$

He next analyzed the probability of the phone being answered exactly twice. He again identified the three distinct paths listed below.

Path	Day		
	1	2	3
1	not answered	answered	answered
2	answered	not answered	answered
3	answered	answered	not answered

Table 4.1.2: Two calls answered out of three – Mutually exclusive paths

Again, each of these mutually exclusive events has a probability of 0.125.

$$P(X=2) = 0.125 + 0.125 + 0.125$$
$$= 3(0.125)$$
$$= 0.375$$

In summary, he noted that if none of the three calls were answered, there was cause for concern. However, he would need more convincing evidence before taking action. He also observed that the likelihood of one call or two calls being answered was exactly the same. Thus if only one out of three had been answered which is below the corporate standard, there would not yet be cause for concern.

4.1.1 Four days at 8:02 a.m. with a standard of 50%

Mr. Smith decided to use the same approach to explore a four-day series of calls. The extra day doubles the size of the tree, because now there are 16 possible sequences. They are depicted in Figure 4.1.2. Each of the 16 sequences is equally likely because the outcomes "answer" and "no answer" both have a probability 0.5. The top path corresponds to the call being answered every day. Its probability of occurrence is $(0.5)^4$ or 0.0625. At the other extreme, no calls answered, also has a probability of 0.0625.

Mr. Smith thought about the possibility of four days in a row with no answer. If the 50% standard were generally being met, this would not likely occur. It has only a 0.0625 probability. If this were to happen, he was thinking that he would call in the help line manager and ask for an explanation.

There is no single criterion to determine if there is enough evidence that a problem exists. The alternative interpretation is the results are simply random fluctuations and no problem exists.

Three common criteria that are considered are 0.1, 0.05, and 0.01. The first criterion says that if there is less than a 10% chance that this poor outcome would occur randomly, the manager should take a closer look and take action to correct a problem. The 0.05 criterion is a higher standard of evidence. This says that if there is less than a 5% or 1-in-20 chance that this poor outcome would occur randomly, the manager should take a closer look and take action. Similarly, a 0.01 criterion involves a 1-in-100 chance of occurrence.

The decision as to what criterion to use involves balancing two points of view with regard to taking action or not taking action to correct a potential problem.

- What is involved in taking action to correct a possible problem? Is there a significant cost associated with taking action? Could taking action when no problem really exists cause harm?

- What are the consequences of waiting before taking action if there is, in fact, a problem? What is the cost of gathering more data before making a decision to act?

Chapter 4 Binomial and Geometric Distributions

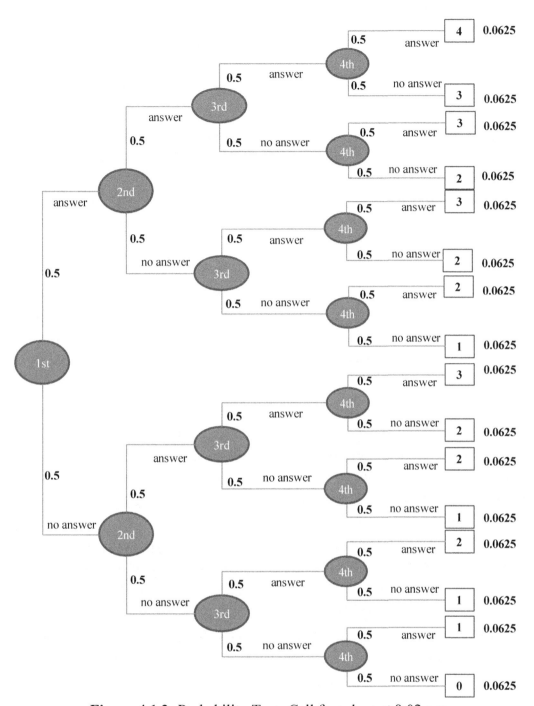

Figure 4.1.2: Probability Tree: Call four days at 8:02 a.m.

If taking action has major costs and consequences, such as firing someone, then the 0.01 criterion might be most appropriate.

1. Describe another situation in which taking action to correct a perceived problem can cause harm if the problem does not really exist.

2. Describe a situation in which delaying action to correct a problem can cause harm.

A scenario of no answers in four days falls between the 0.10 and 0.05 criteria. If Mr. Smith were to use a 5 day sample, the probability of no answer is $(0.5)^5$ which is 0.03. This would definitely cause him to act to correct a perceived problem of not answering the help line at 8:02 am.

Now let's look at some of the other possible outcomes of the phone check to see if call answerers are ready by 8:02 am. If the 50% standard was generally met, then two out of four calls answered would exactly match the standard. If only one out of four was answered, it would be below the standard. However, Mr. Smith was unsure of how strong an indication of a problem this would be. There were four different sequences that involved one answered call. Thus, the probability of this happening was simply

$$P(X=1) = 4(0.0625)$$
$$= 0.25$$

A poor response that has a one-in-four chance of random occurrence should not produce any immediate concerns, nor cause him to act.

Mr. Smith looked back at the process he had been using to perform the calculations. He liked the pictorial representation but knew that when the number of days increased to five or more, it would not be possible to contain the picture on one page. He realized that his approach seemed to involve a standard three-step process.

3. Determine the probability of a specific sequence of events

4. Count the number of different paths, each with the same probability that produce the same total number of answered calls.

5. Multiply the path probability by the number of paths with the same total of positive outcomes.

Mr. Smith already knew how to do step one. It involved just multiplying the individual event probabilities. He wondered if there was a simple formula to help count the different paths. His assistant Sorel Ward recalled a combination formula she had learned in high school and thought it might help. She reasoned as follows.

"There are four paths that result in one answered call. These four paths correspond to the four possible different days. In the language of selecting combinations this is equivalent to asking how many ways are there of picking one (day) out of four (days). This is called four choose one and is represented as $_4C_1$."

We have used the capital letter X, to represent the general random variable. We use the lower case x to represent a specific value of the random variable in the next equation. The general formula for combinations for choosing x items out of n possibilities is

$$_nC_x = \frac{n!}{(n-x)!(x)!}$$

She then considered the case in which calls were answered two days out of the four.

$$_4C_2 = \frac{4!}{(4-2)!(2)!}$$
$$= \frac{24}{(2)(2)}$$
$$= 6$$

The number of paths is six, which is the same as the value determined applying the combinations formula. Her reasoning seemed solid and the numbers matched.

4.1.2 Four Days at 8:05 a.m. with a Standard of 70%

Mr. Smith decided to use this same approach to determine the probability distribution of outcomes for calls placed at 8:05 a.m. The corporate standard was to answer these calls 70% of the time. He started with a probability tree as illustrated in Figure 4.1.3. The layout of sequences in this tree is exactly the same as before. The number of paths that result in one call being answered is again four. The number of paths that result in two calls being answered is again six. The differences in the tree are the probabilities associated with each branch. As a result the end probability of each sequence of four days is different than in Figure 4.1.2. More interestingly, the probability of each sequence is not the same.

In the previous probability tree, the 50% standard means that answer and no answer are equally likely random events. Now the probability of receiving an answer is 0.7 and of receiving no answer is 0.3. Let's look at the two extreme paths of the tree. The top path corresponds to all four calls being answered. It has a probability of occurrence that is $(0.7)^4$ or 0.2401. The bottom path corresponds to no calls being answered. It has a probability of $(0.3)^4$ or 0.0081. If no calls were answered, Mr. Smith would be sure that the call center was not generally meeting the 70% standard. He would act immediately to correct the problem.

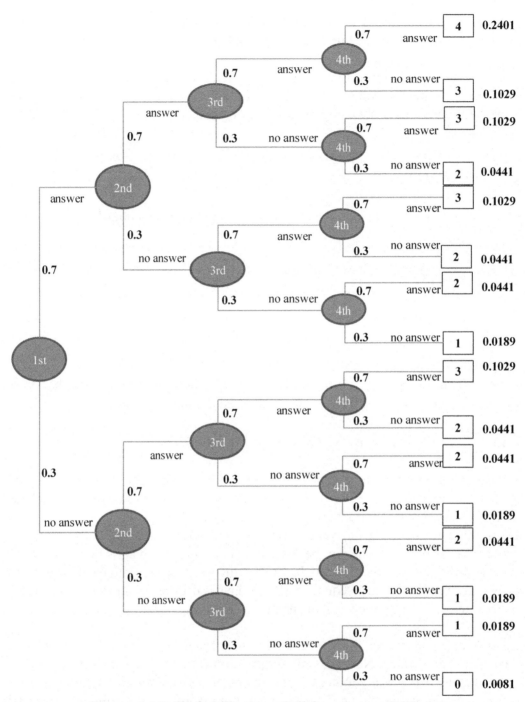

Figure 4.1.3: Probability tree: Call four days at 8:05 a.m.

Mr. Smith looked closely at the probability tree and noticed that specific probabilities occurred multiple times. For example, each sequence that resulted in one call being answered had the same probability of 0.0189. Because of the commutative law of multiplication, the probability of each path is the same. In each case there is one call answered and three calls not answered. The probability of one call being answered is just 0.7. The probability that three calls are not answered is 0.3 raised to the third power.

$$P(\text{one answered call and three unanswered calls}) = (0.7)(0.3)(0.3)(0.3)$$
$$= (0.3)(0.7)(0.3)(0.3)$$
$$= (0.3)(0.3)(0.7)(0.3)$$
$$= (0.3)(0.3)(0.3)(0.7)$$
$$= 0.0189$$

Now, because there are four different paths that result in exactly one of four calls being answered the total probability is

$$4(0.0189) = 0.0756 \quad \text{or} \quad \text{about one in 13.}$$

This probability is below the 10% criterion for random fluctuation. Mr. Smith would, therefore, begin looking more closely at the starting time performance of the call center.

This calculation paralleled the three step approach described earlier.

6. Determine the probability of a specific sequence of events.

7. Count the number of different paths each with the same probability that produce the same total number of answered calls.

8. Multiply the path probability by the number of paths with the same total of positive outcomes.

Mr. Smith repeated the process for all possible values of x, the number of calls answered. He organized the results in Table 4.1.3. A total of three calls out of four answered is the most likely occurrence if the 70% standard was generally being met. Answering two out of four is below the standard. However, it would not be an uncommon occurrence and Mr. Smith would not take action if that happened.

Number of calls answered	Number of calls not answered	Path probability	Number of equally likely paths	Total probability
0	4	$(0.7)^0 (0.3)^4 = 0.0081$	$_4C_0 = 1$	0.0081
1	3	$(0.7)^1 (0.3)^3 = 0.0189$	$_4C_1 = 4$	0.0756
2	2	$(0.7)^2 (0.3)^2 = 0.0441$	$_4C_2 = 6$	0.2646
3	1	$(0.7)^3 (0.3)^1 = 0.1029$	$_4C_3 = 4$	0.4116
4	0	$(0.7)^4 (0.3)^0 = 0.2401$	$_4C_4 = 1$	0.2401
x	$4-x$	$(0.7)^x (0.3)^{4-x}$	$_4C_x = \dfrac{4!}{(4-x)!(x)!}$	$\dfrac{4!}{(4-x)!(x)!}(0.7)^x (0.3)^{4-x}$

Table 4.1.3: Four days at 8:05 a.m. - Probability distribution of the number of calls answered

Based on the above analysis, Mr. Smith, with Sorel Ward's help, believed he could write a general formula to calculate $P(X = x)$ as follows:

$$P(X=x) = \left(_4C_x\right)(0.7)^x(0.3)^{4-x}$$

Let's review this formula. The first part is the number of combinations (sequences) that lead to the same exact number of answered calls, x. It is the number of possible combinations of a collection of four items, chosen x at a time. The second term represents the probability of exactly x answered calls and $(n-x)$ unanswered calls. Because of the commutative product law, this term is exactly the same irrespective of the daily sequence of answered and unanswered calls as long as the total answered is the same x. In each case 0.7 is raised to the power of x, the number of answered calls. And 0.3 is raised to the power of $(4-x)$, the number of unanswered calls.

4.1.3 Four days at 8:10a.m. with a standard of 90%

Mr. Smith is now interested in calculating the probabilities for four calls made at 8:10 a.m. when the corporate standard is 90%. This will guide his thinking as to when he should take action and when he should wait and gather more data. In looking at the above formula, the only changes that need to be made are to replace the 0.7 and its complement 0.3 with the new standard of 0.9 and its complement 0.1. He does not feel a need to build a tree to confirm his reasoning. He asked Sorel Ward to complete the table below.

Number of calls answered	Number of calls not answered	Path probability	Number of equally likely paths	Total probability
0	4	$(0.9)^0(0.1)^4 = 0.0001$	$_4C_0 = 1$	0.0001
1	3		$_4C_1 = 4$	
2	2		$_4C_2 = 6$	
3	1		$_4C_3 = 4$	
4	0		$_4C_0 = 1$	
x	$4-x$	$(0.9)^x(0.1)^{4-x}$	$_4C_x = \dfrac{4!}{(4-x)!(x)!}$	$\dfrac{4!}{(4-x)!(x)!}(0.9)^x(0.1)^{4-x}$

Table 4.1.4: Four days at 8:10 a.m. - Probability distribution of the number of calls answered

4.1.4 Five Days for Different Times and with Different Standards

Mr. Smith was thinking that all tests carried out by AGB should last five days. However, Koala has different standards for 8:02, 8:05, and 8:10 a.m. As a result, he plans to use different criterion for determining whether or not there is strong evidence that a particular standard is not being met. Mr. Smith asked Sorel Ward to apply the binomial distribution for each standard. She needed to generalize the equation that was developed earlier. She reasoned as follows:

- Each time period has a different standard that can be represented by p, the probability the call is answered.

- The probability that a call is not answered is just the complement of this probability. It equals $(1 - p)$.

- In a five day test, the probability that exactly x calls are answered should therefore be:

$$P(X = x) = (_5C_x)(p)^x (1-p)^{5-x}.$$

The first part is the number of combinations (sequences) that lead to the same exact number of answered calls, x. It is the combination of five choose x. The second term represents the probability of exactly x answered calls and $(n - x)$ unanswered calls. Because of the commutative product law, this term is exactly the same for any value of p irrespective of the daily sequence of answered and unanswered calls as long as the total answered is the same x. In each case p is raised to the power of x, the number of answered calls. And $(1 - p)$ is raised to the power of $(5 - x)$ the number of unanswered calls. Sorel used this formula to complete Table 4.1.5.

Number of calls answered	Five-day test of calls answered Time of call and Koala Foods standard		
	8:02 a.m. $p = 0.5$	8:05 a.m. $p = 0.7$	8:10 a.m. $p = 0.9$
x	$P(X = x)$	$P(X = x)$	$P(X = x)$
0	0.031	0.002	0.000
1	0.156	0.028	0.000
2	0.313	0.132	0.008
3	0.313	0.309	0.073
4	0.156	0.360	0.328
5	0.031	0.168	0.590

Table 4.1.5: Five-day test - Probability distribution of the number of calls answered

Mr. Smith reviewed the table and discussed with Sorel the circumstances in which he would take action.

9. Assume AGB conducts a five-day test at 8:02 a.m. Under what circumstances should Mr. Smith take corrective action?

10. Assume AGB conducted a five-day test at 8:02 a.m. Under what circumstances should Mr. Smith be seriously concerned but delay taking action until more data comes in?

11. Assume AGB conducts a five-day test at 8:05 a.m. Under what circumstances should Mr. Smith take corrective action?

12. Assume AGB conducted a five-day test at 8:05 a.m. Under what circumstances should Mr. Smith being seriously concerned but delay taking action until more data comes in?

13. Assume AGB conducts a five-day test at 8:10 a.m. Under what circumstances should Mr. Smith take corrective action?

14. Assume AGB conducted a five-day test at 8:10 a.m. Under what circumstances should Mr. Smith being seriously concerned but delay taking action until more data comes in?

Section 4.2 Binomial Distributions: Expected Value

In Chapter 1, you experimented with flipping a fair coin numerous times. Now consider just two flips of a fair coin. The number of heads can take on the values 0, 1, and 2. Using the binomial distribution, the corresponding probabilities are 0.25 (TT), 0.50 (TH or HT), and 0.25 (HH). The *expected value* of any random variable is simply a weighted sum. First, you multiply each of the possible number of successes by the probability that it occurs. Then, you add those products together. Thus, the expected number of successes for two flips can be computed as shown below.

$$E(X) = \sum x \cdot P(X = x)$$
$$= (0)(0.25) + (1)(0.50) + (2)(0.25)$$
$$= 1 \text{ head}$$

We use the capital letter X, to represent the general random variable. We use the lower case x to represent a specific value of the random variable. Thus, on the left hand side of the above equation we use the capital X. When we take the weighted sum of the possible values, we use the lower case x.

There is a 0.5 likelihood that when flipping the coin twice, you will have one head out of two, which is the expected value. There is also a 0.5 probability of observing some other value. Now consider the case of five days of calling Koala at 8:02 a.m. We calculated the probabilities when $p = 0.5$ in the previous section.

$P(X = 0) = 0.031$
$P(X = 1) = 0.156$
$P(X = 2) = 0.313$
$P(X = 3) = 0.313$
$P(X = 4) = 0.156$
$P(X = 5) = 0.031$

In the above equations we use the capital X. For example, the term $P(X = 2)$ refers to the probability that the random variable X equals the specific value two.

To determine the expected value, we again multiply each of the possible number of successes in a week by the probability that it occurs. Then, we add those products together. Thus, the expected number of calls answered per week can be calculated as shown.

$$E(X) = \sum x \cdot P(X = x)$$
$$= (0)(0.031) + (1)(0.156) + (2)(0.313) + (3)(0.313) + (4)(0.156) + (5)(0.031)$$
$$= 2.5 \text{ calls answered}$$

The expected number of answered calls is 2.5. Obviously, it is not possible to get this value since the number of times the call was answered must be an integer value. If that is the case, what does the term expected value of this random variable really suggest? The answer relates to understanding what happens as you repeat a random experiment over and over again. The calculated average of larger and larger samples will be very close to the expected value.

Above we calculated the expected value by using the basic weighted formula for expected value. The general formula for the expected for the binomial distribution is given below.

Let n = number or repetitions
p = probability of success

$E(X) = np$

This formula makes intuitive sense. For example, if n = 10 tries and p = 0.5, the expected value is five successes. In other words the long-term average will be five successes out of 10 tries. This formula is not limited to p values of 0.5. For example, the Koala standard for 8:05 a.m. is that 70% of the calls should be answered. Using this higher standard, the expected value of the number of calls answered for a five day week is just five times 0.7 or 3.5 calls.

Chapter 4 — Binomial and Geometric Distributions

Section 4.3 *The Lancer* – What if We Publish an Incomplete Paper?

In Chapter 1, we also explored the challenge of publishing the school newspaper, *The Lancer*, on time. Recall there were ten different writers, five editorial writers and five reporters. Even if each of the writers were 95% reliable in meeting the deadline, there was only a 60% chance the paper would be published on time. This was determined by using the multiplication rule for independent events: 0.95 raised to the 10th power is only 0.60.

Mr. Mitchell is committed to publishing the paper on time. He is considering publishing the paper even if all of the articles have not been submitted. He would like the paper to be on time at least 95% of the time. Mr. Mitchell is considering going to press even if only 9 articles have been submitted. He wonders if this will enable him to meet his goal of at least 95% of the time. Mr. Mitchell is considering increasing the number of writers above 10. He plans to evaluate the impact of changing n, the number of writers.

In the previous section we developed the binomial distribution for different numbers of repetitions. We considered three days, four days and five days. We now generalize the distribution for all values of p, the probability of success, and all values of n, the number of repetitions.

Let x = number of on-time submissions,
 n = number of repetitions, and
 p = probability of success each time.

$$P(X = x) = \left({}_nC_x\right)(p)^x (1-p)^{n-x}$$

The first term is the number of different sequences that will result in exactly x successes and $(n - x)$ failures. The rest of the equation is the probability any individual sequence has exactly x successes and $(n - x)$ failures.

Now let's return to the specifics of Mr. Mitchell's decision. In terms of probability, we can express the probability of on time publication with a policy of publishing with only nine articles as follows:

$$P(X \geq 9) = P(X = 9) + P(X = 10)$$

This includes two terms. If nine or 10 articles are submitted on time, the paper is published on time.

Using the binomial formula, we obtain the following result.

$$P(X \geq 9) = P(X = 9) + P(X = 10)$$
$$= (_{10}C_9)(0.95)^9(1-0.95)^1 + (_{10}C_{10})(0.95)^{10}(1-0.95)^0$$
$$\approx 0.315 + 0.599$$
$$= 0.914$$

This policy is close to his goal. If he were to use this policy, sometimes the paper would have 10 articles and other times only nine. He is interested in knowing the proportion of editions published on time that are complete. This is an example of *conditional probability*: What is the probability that the issue is complete, *given* that it goes to press on time? The statement, "It goes to press on time," is a given *condition*. This is the source of the term *conditional* probability. To answer this question we write the following conditional probability statement.

$$P(X = 10 \text{ given the paper is published on time}) = P(X = 10 | X \geq 9)$$

In words this means that, given there were nine or more articles submitted on, what is the probability that were in fact exactly ten articles submitted on time. The conditional probability formula is

$$P(A|B) = \frac{P(A \cap B)}{P(B)}$$

The denominator represents the probability of the given event, that the paper is published on time in this case.

$$P(X = 10 | X \geq 9) = \frac{P(X = 10 \cap X \geq 9)}{P(X \geq 9)}$$

The numerator $P(X = 10 \cap X \geq 9)$ contains the intersection of two events with regard to X. However, the only way that X can *both* equal 10 and be greater than or equal to nine is if X equals 10. Therefore,

$$P(X = 10 | X \geq 9) = \frac{P(X = 10)}{P(X \geq 9)}$$
$$\approx \frac{0.599}{0.914}$$
$$\approx 0.655$$

When the paper is published on time, there is a little less than two-thirds likelihood that paper is complete with all 10 articles.

4.3.1 Meeting the 95% Goal for Publishing on Time

The above policy would not enable Mr. Mitchell to meet his 95% on time goal. He therefore considers expanding the policy and allowing the paper to be published with as few as 8 articles. We now need also to find the probability that exactly eight articles are available on time.

$$P(X=8) = (_{10}C_8)(0.95)^8(1-0.95)^2$$
$$= 0.075$$

Now, the probability of eight or more articles being available on time is

$$P(X \geq 8) = P(X=8) + P(X=9) + P(X=10)$$
$$\approx 0.075 + 0.315 + 0.599$$
$$= 0.988$$

Thus if Mr. Mitchell go to press with just eight articles, the paper will be published on time almost 99% of the time.

When the paper is published on time, it will have eight, nine, or 10 articles. He is now concerned how frequently the on time paper will have as few as eight articles. This is another conditional probability question. To answer it, we write the following conditional probability statement:

$P(X = 8 \mid$ the paper is published on time$) = P(X = 8 \mid X \geq 8)$

Applying the conditional probability formula, we have

$$P(X=8 \mid X \geq 8) = \frac{P(X=8 \cap X \geq 8)}{P(X \geq 8)}.$$

Now, the only way that X can be *both* greater than or equal to 8 and equal to eight is if it is exactly equal to eight. In other words, $P(X=8 \cap X \geq 8) = P(X=8)$. Thus,

$$P(X=8 \mid X \geq 8) = \frac{P(X=8)}{P(X \geq 8)}$$
$$\approx \frac{0.075}{0.988}$$
$$\approx 0.076$$

Approximately, one out of every 13 on-time editions will have only eight articles, because $1/0.076 \approx 13.2$.

Chapter 4 — Binomial and Geometric Distributions

4.3.2 Twelve Writers – But Less Reliable

This year Mr. Mitchell had an unusually large group of students interested in working on *The Lancer*. He selected the best 12 of the group. However, this group seems to him to be much less reliable than past groups. He estimates, based on past experience, that each of the individuals in this group will meet the paper deadline 80% of the time. Recall that the expected value of the binomial distribution is $E(X) = np$, where n is the number of possibilities (writers, in this case), and p is the probability of a success (meaning the writer meets the deadline in this case). So, the expected value of the number of writers meeting the deadline in this case is

$$E(X) = (12)(0.8) = 9.6 \text{ writers who met their deadline.}$$

Thus on average there would be fewer than 10 articles ready on time for publication. He once again is ready to consider publishing the paper with fewer than ten articles. To explore his options, he decides to use a cumulative table of the Binomial Distribution instead of performing calculations. The cumulative distribution sums all the probabilities up to and including a specific value of X. It is written as $P(X \leq x)$. For example if $x = 8$, the cumulative probability is the probability that the random variable takes on any value from zero up to eight. It involves summing up the probabilities as in the equation below.

$$P(X \leq 8) = P(X=0) + P(X=1) + P(X=2) + P(X=3) + P(X=4) + P(X=5) + P(X=6) + P(X=7) + P(X=8)$$
$$\approx 0 + 0 + 0 + 0.0005 + 0.0033 + 0.0155 + 0.0532 + 0.1329$$
$$= 0.2054$$

With graphing calculators, it is not necessary to perform each step of this calculation. The cumulative distribution is a calculator function: binomcdf(. Table 4.3.1 lists $P(X \leq x)$ for every possible value of x from zero to 12 for a range of values for p. For example, if $p = 0.8$, the probability that the number of successes will be eight or fewer is 0.205. If $p = 0.8$, the probability that the number of successes will be 10 or fewer is 0.725. Mr. Mitchell is, however, more interested in probability statements of the form $P(X \geq x)$. This can be determined from the table by using the principle of the complement. Thus, for example, the complement of nine or more is eight or fewer. Applying the principle of the complement, the probability of nine or more is equal to one minus the probability of eight or less.

$$P(X \geq 9) = 1 - P(X \leq 8)$$
$$\approx 1 - 0.205$$
$$= 0.795$$

Mr. Mitchell found the table to be very helpful. He asked one of the students to calculate this probability by determining the probability of each individual outcome.

$$P(X \geq 9) = P(X=9) + P(X=10) + P(X=11) + P(X=12)$$
$$= (_{12}C_9)(0.8)^9(0.2)^3 + (_{12}C_{10})(0.8)^{10}(0.2)^2 + (_{12}C_{11})(0.8)^{11}(0.2)^1 + (_{12}C_{12})(0.8)^{12}(0.2)^0$$

1. What are the individual values and the total?

After watching the student go through each calculation, Mr. Mitchell was glad to have this simple lookup table.

In general, when X and x must be integers, $P[X \geq (x+1)] = 1 - P(X \leq x)$.

	\multicolumn{4}{c}{p values}			
	0.75	0.8	0.85	0.9
$X \leq 0$	0.000	0.000	0.000	0.000
$X \leq 1$	0.000	0.000	0.000	0.000
$X \leq 2$	0.000	0.000	0.000	0.000
$X \leq 3$	0.000	0.000	0.000	0.000
$X \leq 4$	0.003	0.001	0.000	0.000
$X \leq 5$	0.014	0.004	0.001	0.000
$X \leq 6$	0.054	0.019	0.005	0.001
$X \leq 7$	0.158	0.073	0.024	0.004
$X \leq 8$	0.351	0.205	0.092	0.026
$X \leq 9$	0.609	0.442	0.264	0.111
$X \leq 10$	0.842	0.725	0.557	0.341
$X \leq 11$	0.968	0.931	0.858	0.718
$X \leq 12$	1.000	1.000	1.000	1.000

Table 4.3.1: Binomial cumulative distribution for $n = 12$

Thus, to explore his options, Mr. Mitchell is looking for the value of x such that

$$P[X \geq (x+1)] \geq 0.95.$$

Using the principle of the complement, we are looking for the largest value of X such that

$$P(X \leq x) \leq 0.05.$$

Mr. Mitchell looked at the cumulative probabilities for both X less than or equal to seven and X less than or equal to six. He found the following

$$P(X \leq 7) \approx 0.073 \quad \rightarrow \quad P(X \geq 8) \approx 1 - 0.073 = 0.927$$
$$P(X \leq 6) \approx 0.019 \quad \rightarrow \quad P(X \geq 7) \approx 1 - 0.019 = 0.981$$

Thus if he wants the Lancer to be published on time at least 95% of the time, he will have to settle for a policy of publishing with seven or more articles.

2. Do you think the Lancer should be published with seven articles when the basic design was made for 10 articles? What standard would you recommend and how often would you be able to meet that standard?

Mr. Mitchell was concerned that the paper might seem skimpy if there were only seven articles. He was thinking about a strategy that might help fill in the gap. One thought that came to mind is that sometimes there might be a surplus of articles for an on-time edition. There are twelve writers and only 10 articles can be published in an issue. In that case, he could withhold one or more of these articles for a subsequent edition. Even though this year's writers were not as reliable as in the past, it was possible that 11 or even 12 of the writers might submit articles on time. He was therefore interested in determining the likelihood of this happening. He looked at the table for the value of r equal to 10 and applied the principle of the complement.

$$P(X \geq 11) = 1 - P(X \leq 10)$$
$$= 1 - 0.725$$
$$= 0.275$$

He found that more than 27% of the time, he would have 11 or more articles. He could then publish a full-edition paper with ten articles. He would hold onto the one or two extra articles to use them later in the year if fewer than ten articles are submitted.

Mr. Mitchell estimated that the reliability of the writers was 0.8 based on experience with similar teams in the past. He had some concern about his estimate. He also knows that if he aggressively monitors their work habits, he may be able to increase their reliability.

3. Assume the individual reliability is only 0.75. What should he use for the minimum number of articles so he can meet the 95% goal? What policy would you recommend?

If he supervises closely, he can increase the reliability to 85% after a couple of issues of the Lancer.

4. Assume the individual on time reliability reaches 0.85. What standard should he use for the minimum number of articles so he can meet the 95% goal?

Mr. Mitchell believes that by the second semester, it may be possible to increase the reliability to 90%. He was wondering what would be the benefit of his efforts to increase reliability.

5. Assume the individual on time reliability reaches 0.90. What standard should he use for the minimum number of articles so he can meet the 95% goal?

Chapter 4 — Binomial and Geometric Distributions

Section 4.4 Worker Absenteeism at BT Auto Industries

BT Auto Industries is a small manufacturer based in Detroit, Michigan. It recently received several large orders and needs 25 workers on the job each day. Management wants to know how many spare workers to hire during the flu season. The last time they carried out a study, they simulated 10 weeks of flu season, 50 days. The process of simulation was time consuming. Kenneth Young is the chief industrial engineer. He knows that the Binomial Distribution can be used to perform the analysis much more quickly and effectively.
He knows if he does not have spare workers, most days he will be short workers. The likelihood that 25 workers out of 25 show up during flu season is easily calculated with the binomial probability density function.

$$P(X=25) = (_{25}C_{25})(0.9)^{25}(1-0.9)^{0}$$
$$\approx 0.072$$

It is worse than he thought. There is only a 7.2% chance everyone will show up for work. That means on average 13 out of every 14 days he will be short workers. He next considers having just 1 spare worker available every day. This worker may also not show up if he has the flu. He will have enough workers if at least 25 of these workers and spares come to work.

$$P(X \geq 25) = P(X=25) + P(X=26)$$
$$= (_{26}C_{25})(0.9)^{25}(1-0.9)^{1} + (_{26}C_{26})(0.9)^{26}(1-0.9)^{0}$$
$$\approx 0.187 + 0.065$$
$$= 0.252$$

This is a significant improvement but nowhere near the 95% goal. He also sees that the calculations will continue to get more complex as he adds spare workers. He will need to add together more and more terms. His trusty TI calculator can simplify this if he uses the concept of the complement and the cumulative distribution function of the calculator. The complement of the event that 25 or more workers show up is that 24 or fewer come to work. Thus, the last calculation can be performed as follows.

$$P(X \geq 25) = 1 - P(X \leq 24)$$
$$= 1 - \text{binomcdf}(26, 0.9, 24)$$
$$\approx 1 - 0.748$$
$$= 0.252$$

The first parameter within the parenthesis is n, the number of workers including spares. In this case it is 26. The second parameter is 0.9, the probability that a worker shows up. The third parameter is the value of interest, 24 workers or less. He was also interested in the average number of workers that would come in for each policy. Recall that the expected value is simply $n \times p$. With 25 workers, the average number that shows up is 22.5.

Kenneth Young asked his assistant, Jay Trust, to continue to evaluate different numbers of spare workers by completing Table 4.4.1.

Spare workers	Total number of workers	$P(X \leq 24)$	$P(X \geq 25)$	Average number of workers who show up ($n \times p$)
0	25	0.928	0.072	22.5
1	26	0.749	0.251	23.4
2	27			
3	28			
4	29			
5	30			
6	31			
7	32			

Table 4.4.1: Spare workers and probability of having at least 25 workers

1. How many spare workers are needed to reach the goal of 95% of the time there are at least 25 workers in the manufacturing plant?

2. On average how many workers will show up for work under this policy? How many will not actually be used?

3. Under this policy, what proportion of days will they still be short exactly one worker? On average how many days during 50-day flu season will they be short exactly one worker?

4. Under this policy, what proportion of days will they still be short two or more workers?

5. The company is considering a slightly lower standard than 95%? What would the standard have to be in order to reduce the number of spare workers by one?

6. The Centers for Disease Control and Prevention (CDC) has been tracking the flu season. It has been worse than in past years. During the peak three weeks of the flu season, the CDC forecasts that 15% of workers will be out sick on any random day. How many spare workers should they schedule for the peak flu season to meet the 25-worker goal 95% of the time?

Section 4.5 Hall of Fame Blogs and Geometric Distribution

Thaddeus Boggs is CEO of Hall of Fame Blogs (HOF Blogs). The organization's website hosts bloggers who write humorous observations about recent events in the world of sports. He has recruited 30 famous athletes, each of whom is known for his or her sense of humor and has solid writing skills. These athletes will post to their blogs on a weekly basis. They will share in the advertising revenue generated by the blog site. In order to be part of this group, each athlete committed to publishing new material to their blogs, on average, once every three weeks.

Mr. Boggs wants to be sure he has enough posts each week for sports fans to keep coming back to the site. He is hoping that with 30 bloggers, he will have plenty of material.

1. On average how many blog posts will be published each week on HOF Blogs if the athletes meet their minimum commitments?

Thad is planning on spreading out the blog posts over the entire week. This will give readers a reason to check out the site every day for the latest material.

2. What is the probability that he will have at least seven blogs available to post during the first week?

Ideally, Mr. Boggs would like 95% of the weeks to have enough bloggers contributing new material to average at least one post per day.

3. Assume each blogger will post, on average, once every three weeks. What is the minimum number bloggers HOF Blogs needs to achieve the goal of having at least seven new posts each week 95% of the time? Is it 30? 31? 32? 33? 34? 35?

After four weeks, Thad decided to see who had contributed and who had not. Thad thought he had received strong commitments from all of his star athletes. He was surprised to find that there were six athletes who had not yet posted to their blog. As a result, Mr. Boggs recruited Daniel Robinson, a graduate of the US Naval Academy, to track participation of the individual bloggers. It is his responsibility to contact individuals who seem to be not keeping their commitments. The Commander, as Mr. Robinson was nicknamed, would discuss the possibility of dropping those individuals from the program.

The Commander's education at Annapolis included basic courses in probability and statistics. He had studied both the binomial and geometric distributions. He felt the geometric distribution could be applied to this context. Below are the key assumptions for the **geometric distribution**.
- The probability of success on any individual trial is p.
- The chance of success on any individual trial is independent of what happened on every other trial.

Chapter 4 — Binomial and Geometric Distributions

These assumptions are the same as with the binomial distribution. The geometric differs from the binomial with regard to the definition of the random variable that is being tracked. In the binomial, the random variable X is the number of successes in n trials. It takes on whole number values from zero up to n. In contrast, in the geometric distribution, the random variable Y corresponds to the number of trials up to and including the first success. The lowest possible value of this random variable is one. It is impossible to have a success without the first trying. Conversely, there is no theoretical limit to how many trials it might take to achieve the first success. Thus, the upper bound on Y is infinity.

Let S_i = success in posting to the blog in Week i, and
F_i = failure to post to the blog in Week i.

Thus, $P(S_i) = \dfrac{1}{3}$ and

$$P(F_i) = 1 - P(S_i) = \dfrac{2}{3}.$$

The Commander recalled that the formula for the geometric distribution was easy to derive. For example, the probability that at an athlete would post on his or her blog in the first week is simply 0.33. This corresponds to the random variable Y having the value one.

$$P(Y=1) = P(S_1)$$
$$= \dfrac{1}{3}$$

If Y were to equal two, that means the athlete did not post the first week but posted the second week.

$$P(Y=2) = P(F_1 \cap S_2)$$
$$= P(F_1) \cdot P(S_2)$$
$$= \left(\dfrac{2}{3}\right) \cdot \left(\dfrac{1}{3}\right)$$
$$= \dfrac{2}{9}$$
$$\approx 0.22$$

4. Why is it permissible to use the multiplication rule to determine $P(F_1 \cap S_2)$?

The likelihood that an athlete's first post occurred in the third week follows the same logic. The first two weeks, the athlete did not post and on the third week, she posted.

$$P(Y=3) = P(F_1 \cap F_2 \cap S_3)$$
$$= P(F_1) \cdot P(F_2) \cdot P(S_3)$$
$$= \left(\frac{2}{3}\right)^2 \cdot \left(\frac{1}{3}\right)$$
$$= \frac{4}{27}$$
$$\approx 0.15$$

Similarly,

$$P(Y=4) = P(F_1 \cap F_2 \cap F_3 \cap S_4)$$
$$= P(F_1) \cdot P(F_2) \cdot P(F_3) \cdot P(S_4)$$
$$= \left(\frac{2}{3}\right)^3 \cdot \left(\frac{1}{3}\right)$$
$$= \frac{8}{81}$$
$$\approx 0.10$$

With the above analysis, Thad was able to determine, the probability that an athlete would post his first blog on or before the fourth week. That probability is 0.8

$$P(Y \leq 4) = P(Y=1) + P(Y=2) + P(Y=3) + P(Y=4)$$
$$\approx 0.33 + 0.22 + 0.15 + 0.10$$
$$= 0.80$$

The complement of this event corresponds to the athlete who has not yet posted her first blog. That athlete's first post will therefore occur during the fifth week or later.

$$P(Y \geq 5) = P(Y > 4)$$
$$= 1 - P(Y \leq 4)$$
$$\approx 1 - 0.80$$
$$= 0.20$$

This means that, strictly due to randomness, there is a one-in-five chance that by week four an athlete would not yet have posted anything on his blog. It is therefore not surprising, that six out of the 30 athletes, 20 percent had not yet posted their first entries, even though they were committed to the business.

The Commander also realized there was a more direct way to calculate the probability of the complementary event. The following equivalent relationship helps with the calculation.

> $Y \geq 5$ if and only if there are no successes in the first four trials.

$$P(Y \geq 5) = P(F_1 \cap F_2 \cap F_3 \cap F_4)$$
$$= \left(\frac{2}{3}\right)^4$$
$$= \frac{16}{81}$$
$$\approx 0.20$$

This formula can be generalized as the following.

$$P(Y > y) = P(\text{no successes in first } y \text{ trials})$$
$$= P(F_1 \cap F_2 \cap F_3 \cap \ldots \cap F_y)$$
$$= \left(\frac{2}{3}\right)^y$$

Thad wondered at what point he should confront an athlete who has not yet published his or her first blog post. The Commander suggested waiting until the probability of this occurring purely by chance was less than one in 30, or 0.033. This probability is shown in the equation below.

$$P(Y > y) = \left(\frac{2}{3}\right)^y$$
$$\frac{1}{30} = \left(\frac{2}{3}\right)^y$$
$$0.033 \approx \left(\frac{2}{3}\right)^y$$

5. Solve the above equation to determine the value of y.

6. Should we round the value of y down or up? Why?

7. Complete the Table 4.5.1 below.

y	P(Y = y)	P(Y ≤ y)	P(Y > y)
1	0.33	0.33	0.67
2	0.22	0.55	0.45
3	0.15	0.70	0.30
4	0.10	0.80	0.20
5			
6			
7			
8			
9			
10			
11			
12			

Table 4.5.1: Geometric probability distribution with $p = 1/3$

Daniel Robinson thought of another way to make this analysis more concrete. Boggs, a former baseball player, understood data but was not that great with mathematical formulas. The Commander decided to **simulate** the posting pattern of 30 bloggers. For each athlete he generated 15 random numbers. (Electronic random number generators return numbers that are uniformly distributed between zero and one.) Any number less than or equal to 0.333 would represent success in publishing a blog post. Any number greater than 0.333 would represent a failure to post to the blog. Table 4.5.2 presents the simulation of four athletes. This simulation corresponds to the binomial distribution with $n = 15$ weeks and $p = 0.333$ chance of success.

Athlete	Week														
	1	2	3	4	5	6	7	8	9	10	11	12	13	14	15
A	S	F	S	S	F	F	F	F	F	F	F	S	S	F	F
B	F	S	F	F	F	F	F	F	S	F	S	F	F	F	F
C	S	S	F	S	F	S	S	S	S	S	F	S	S	F	F
D	F	F	F	F	S	F	F	F	F	S	F	F	F	F	F

Table 4.5.2: Binomial simulation of four bloggers over 15 weeks

However, the Commander was only interested in the time until the first blog post was published. He reviewed each row and stopped at the first successful post. Anything after the first post was irrelevant with regard to the geometric distribution. Thus the simulation presented in Table 4.5.2 was converted into Table 4.5.3. Athletes A and C published on their posts in the first week. Athlete B's first post was in week two. Athlete C published her first post in week five.

Athlete	Week														
	1	2	3	4	5	6	7	8	9	10	11	12	13	14	15
A	S														
B	F	S													
C	S														
D	F	F	F	F	S										

Table 4.5.3: Conversion to geometric simulation of four bloggers until first blog

The Commander reviewed the simulation results in Table 4.5.4. He pointed out that only 21 of the 30 bloggers had published at least one post by the fourth week. Even after six weeks, five athletes had yet to post their first piece. It was not until week 10 that everyone had contributed at least one post.

8. Carry out your own simulation of 30 athletes. Summarize your results as in Table 4.5.4.

Mr. Robinson also talked to Mr. Boggs about his own baseball career. Mr. Boggs had a lifetime batting average close to .333. The Commander asked if Thad could recall times when he batted and did not get a hit eight, 10, or even 12 times in row. Thad acknowledged that over the course of the many 162-game seasons in which he batted more than 600 times, he had several such long streaks each year. It usually did not mean he was in slump. It was simply a part of the random fluctuations around the mean. Long strings of failures occasionally occurred even with the best hitters in baseball.

After discussing the geometric distribution, Thaddeus Boggs and Daniel Robinson decided to implement a two-phase strategy. Any player who had not submitted a blog post during the first five weeks would be contacted by Daniel Robinson. The Commander would talk with the player about his or her interest in continuing with the program. He would encourage the athlete to post new material. However, no other action would be taken. Any player who had not submitted his or her first blog by week eight would be dropped from the program.

9. Explain whether or not you support this plan?

10. What would you recommend as alternative strategy?

Athlete	Week										
	1	2	3	4	5	6	7	8	9	10	11
A	S										
B	F	S									
C	S										
D	F	F	F	F	S						
E	F	F	F	S							
F	F	F	F	F	F	F	F	F	F	S	
G	F	F	F	F	F	F	F	F	S		
H	F	F	S								
I	F	F	F	F	F	F	F	S			
J	S										
K	F	F	F	S							
L	F	F	S								
M	F	F	F	F	F	F	S				
N	F	S									
O	F	F	S								
P	S										
Q	S										
R	F	S									
S	F	F	S								
T	S										
U	S										
V	S										
W	F	S									
X	S										
Y	F	F	F	F	F	S					
Z	F	F	F	F	F	F	F	F	S		
AA	F	S									
AB	F	F	F	F	F	S					
AC	F	F	F	S							
AD	F	F	F	F	F	S					
Totals	9	5	4	3	1	3	1	1	2	1	0
Cumulative	9	14	18	21	22	25	26	27	29	30	30

Table 4.5.4: Geometric simulation of 30 bloggers until first post

Chapter 4   Binomial and Geometric Distributions

Section 4.6 First Unanswered Call - Geometric Distribution – Koala Foods 8:10 a.m.

We will again consider the quality control concern of Koala Foods. In Chapter 1, we investigated the probability that x successes occurred. Success was defined as the call reaching a customer service representative at a specified time. This gave rise to the binomial distribution. In this section, we will investigate the probability that the first time an 8:10 a.m. call is not answered occurs on the rth day that we call. This will help us develop an understanding of a new distribution that describes the first time an event occurs.

This pattern is called the ***geometric distribution***. It arises in numerous contexts. For example, in the pharmaceutical industry, managers are interested in the first compound of many they have tested that seems to address a specific disease. An outcome of interest can be a negative event. NASA managers were concerned with predicting probabilistically the first time they would have a catastrophic event during a shuttle mission.

After many weeks of working with the call center staff, Mr. Smith of Koala Foods is convinced that the 90% standard for 8:10 a.m. is being achieved. Nevertheless, he understands that there is still a 10% chance a call will not be answered. He wonders if AGB were to run a daily 8:10 a.m. test, what would be the first day a call was not answered? He begins by asking and answering a series of questions.

- What is the probability of not reaching a customer service representative at 8:10 a.m. on the first day? This he realizes is just 0.1, the likelihood that a call is not answered.

- What is the probability of reaching a customer service representative on the first day, and not reaching one on the second day? The likelihood that a call is answered the first day is 0.9. The chances of no answer the second day is still, 0.1. The likelihood of this sequence is just the product of these two numbers, 0.9 × 0.1. This equals 0.09.

- Next, what is the probability of reaching a customer service representative at 8:10 a.m. on the first day and second day, but not reaching one on the third day? The likelihood of an answered call on the first day is 0.9. The same probability applies to the second day. The chances of no answer the 3nd day is still 0.1. The likelihood of this sequence is just the product of these three numbers, 0.9 × 0.9 × 0.1. This equals 0.081.

He reviews the questions and answers and is interested in developing a common pattern. Mr. Smith decides to approach the problem like he did with the binomial distribution. There he started by counting the number of answered calls over three days. For this problem, he first develops a table and then a probability tree. In Table 4.5.1, Mr. Smith lays out each sequence discussed above. In Figure 4.5.1 he builds a probability tree but stops at five days of calls.

Chapter 4 — Binomial and Geometric Distributions

Sequence	Day		
	1	2	3
1	not answered		
2	answered	not answered	
3	answered	answered	not answered

Table 4.5.1: Sequence of days until call is not answered for first time

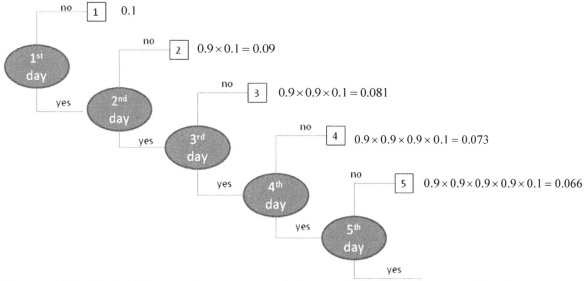

Figure 4.5.1: Probability tree for sequence of days until call is not answered for first time

Mr. Smith compared this table and figure to the earlier ones, Tables 4.1.1 and Figure 4.1.1, He is struck by how different they are. Here are his main observations about the table and figure.

- Looking at the table, he saw no sequences that looked the same. Each sequence was unique. Similarly, no number appeared more than once at the end of the tree.

- There is no zero value in the tree or table.

- He stopped the probability tree at five days but there seems to be no limit to the number of days.

- The probabilities next to each sequence in the probability tree are all different.

After looking this over, Mr. Smith's initial reaction was "*oy vey*." After closer inspection, he thought he saw a simple pattern. In each sequence there were consecutive days of answered calls followed by one unanswered call. For example, consider when the first unanswered call occurs on the fourth day. This means three days of answered calls followed by one day with no answer at 8:10 a.m. The probability is easily calculated by the multiplication rule of probability of independent events. We will let R be the random variable representing the number of the day in which the first unanswered call is observed.

$$P(R=4)=(0.9)(0.9)(0.9)(0.1)$$
$$=(0.9)^3(0.1)$$

He also saw that he could simplify the formula as 0.9 raised to the third power multiplied by 0.1. He repeated this for $r = 5$.

$$P(R=5)=(0.9)(0.9)(0.9)(0.9)(0.1)$$
$$=(0.9)^4(0.1)$$

He then reasoned that this could be generalized for any value R. In order for the first unanswered call to occur on Day r, there must be $(r-1)$ days of answered calls followed by one day of no answer. This is the one and only possible sequence.

$$P(R=r)=(0.9)^{r-1}(0.1)$$

This is the formula for the geometric distribution with the probability of occurrence (an unanswered call) equal to 0.1. He decided to organize the probabilities in Table 4.5.2 and records the probabilities up to $r = 5$ days.

Number of the day that customer service is NOT reached for the first time: r	Probability customer service is reached for the first time on Day r
0	
1	0.1
2	$(0.9)(0.1) = 0.09$
3	$(0.9)^2(0.1) = 0.081$
4	$(0.9)^3(0.1) = 0.073$
5	$(0.9)^4(0.1) = 0.066$
6	
7	
8	
9	
10	

Table 4.5.2: Probabilities of first recording occurring on Day r

1. What is happening to the probability as the number of days increases?

2. If you know the probability, $P(R = r)$, what calculation is required to find $P[R = (r + 1)]$?

3. Why is there no probability value for $r = 0$? Why is this different than the binomial distribution?

4. Use a calculator to complete the above table. What is the lowest value in the table?

5. Can you find a number of days for which the calculator returns a probability of zero?

6. What would a probability of zero mean? Do you believe that the probability is actually zero? Explain.

Mr. Smith was still somewhat puzzled by this distribution. It seems there is no largest value. It goes off to infinity. He is concerned because one of the fundamental concepts in probability theory is that the sum of all the probabilities must equal one. He added up the first five values and the total was 0.411. This is equivalent to the following equation.

$$P(R \le 5) = 0.410$$

He interpreted this to mean, that if he called at 8:10 a.m. every morning, there is a 0.410 probability that the first time the call goes unanswered will occur on or before the fifth day. Thus, there was less than a 50-50 chance that he would notice the first unanswered call before five days have passed.

He then added up the first 10 probabilities. The answer was 0.651. This was higher but still not very close to one.

7. What is the total for the first 20, 25 and 30 values? Are these close to one?

8. Determine for what value of R is the sum total greater than 0.99.

Mr. Smith asked the consultants at CBG if they could explain how the sum of the probabilities could add to one. Dr. Nat Finit reminded Mr. Smith about the concept of the sum of an infinite geometric series when the common ratio is strictly less than one. The probabilities of the geometric distribution form an infinite geometric series.

Let a = value of first term
p = the common ratio multiplier used to calculate each succeeding term

The formula for the infinite sum, S, is

$$S = \frac{a}{1-p}$$

In this problem $a = 0.1$ and $p = 0.9$. The numerator, a, is the probability of no answer and the term p is its complement.

$$\begin{aligned} S &= \frac{0.1}{1-0.9} \\ &= \frac{0.1}{0.1} \\ &= 1 \end{aligned}$$

Thus the *infinite* sum does equal one. Dr. Finit also explained that, for the geometric distribution, a and p will always be complements. Thus, the sum total will always be one for any pair of complementary values a and p.

Mr. Smith found that calculating the cumulative probability was tedious.

$$P(R \leq 5) \approx 0.1 + 0.09 + 0.081 + 0.073 + 0.066$$
$$= 0.410$$

$$P(R \leq 10) \approx 0.1 + 0.09 + 0.081 + 0.073 + 0.066 + 0.059 + 0.053 + 0.048 + 0.043 + 0.039$$
$$= 0.651$$

He wondered if there was an easier way to find these probabilities. Dr. Nat Finit discussed with him a simple trick for determining these probability sums. Dr. Finit described how to use the principle of the complement to indirectly calculate these probabilities.

Let's take the first example $P(R \leq 5)$. If R is less than or equal to five then the first unanswered occurred on or before the fifth day. The complement of this event is that calls were answered on each and every day. There was not one single day in the first five that a call was unanswered at 8:10 a.m.

$$P(\text{calls answered five days in a row}) = (0.9)(0.9)(0.9)(0.9)(0.9)$$
$$= 0.9^5$$
$$\approx 0.590$$

Thus:
$$P(R \leq 5) = 1 - P(\text{calls answered five days in a row})$$
$$\approx 1 - 0.590$$
$$= 0.410$$

The same logic would apply to

$$P(R \leq 10) = 1 - P(\text{calls answered 10 days in a row})$$
$$= 1 - 0.9^{10}$$
$$\approx 1 - 0.349$$
$$= 0.651$$

In general, we can write that

$$P(R \leq r) = 1 - P(\text{calls answered } r \text{ days in a row})$$
$$= 1 - (1 - 0.1)^r$$
$$= 1 - 0.9^r$$

This formula was developed for an example in which the probability of an outcome was 0.1. This formula calculates the cumulative probability of the outcome being equal to or less than r. It is called the cumulative distribution function and abbreviated as cdf. The same logic would apply for any value of p, the probability of the outcome of interest.

$$P(R \leq r) = 1 - P(r \text{ failures in a row})$$
$$= 1 - (1-p)^r$$

9. Use this formula to determine the likelihood for Mr. Smith that the first time a call goes unanswered occurs on or before the 15th day.

10. Determine for the smallest value of R such that the cumulative likelihood of finally having a call go unanswered is greater than 0.5.

Mr. Smith was interested in one last statistic, the expected value of the number of days until the first unanswered call at 8:10 a.m. He wrote down the general expression for expected value and the specific equation for the 8:10 a.m. case.

$$E(R) = \sum r \cdot P(R=r)$$
$$= 1(0.9)^0(0.1) + 2(0.9)^1(0.1) + 3(0.9)^2(0.1) + 4(0.9)^3(0.1) + \ldots + r(0.9)^{r-1}(0.1) + \ldots$$

Mr. Smith did not see an easy way to sum this infinite series. It was a weighted sum of the geometric series. Dr. Nat Finit chimed in. He said that with using derivatives learned in calculus he could derive a simple formula for the expected value. The general formula for the expected value of a geometric distribution with the probability p of an occurrence is just

$$E(R) = \frac{1}{p}.$$

In this context the expected value is 10 days. On average the first time an 8:10 a.m. call is not answered is on the 10th day.

$$E(R) = \frac{1}{0.1}$$
$$= 10 \text{ days}$$

However, the probability the first unanswered call occurs exactly on the 10th day is very small.

$$P(R=10) = (0.9)^9(0.1)$$
$$\approx 0.039$$

Section 4.7 Geometric Distribution–NASA Shuttle Catastrophic Failure

The National Aeronautics and Space Administration (NASA) and the astronauts all recognize and accept the fact that all manned space flight is inherently dangerous. In addition unlike automotive testing, it is not possible to actually test all of the systems under real-world conditions. Instead, they develop policies and procedures for trying to make the vehicles as safe as possible while being able to meet the space mission needs. Typically, they use engineering judgment to estimate the risks associated with space flight.

As work began on development of the space shuttle, the NASA team of engineers and managers collectively estimated that there was a one-in-80 chance of a catastrophic failure. They anticipated flying the shuttle only 50 times. They were therefore optimistic that there would be no catastrophes. Based on this estimate, we will explore a number of related probabilities. We will also compare their initial estimates to the actual record of catastrophes to see if the experience from 1972 to 2011 was in line with their original estimates.

On July 8, 2011 the space shuttle *Atlantis* was launched on the last mission of the shuttle program. We will use the binomial and geometric distributions to summarize the catastrophic risk probabilities for a total of 135 missions instead of the originally planned 50.

The probability that the first failure will be occur on any specific flight is very small. A one in eighty chance equals a probability of 0.0125. For example, the probability that the first failure occurs on the 25th flight is calculated below.

$$P(R=25) = (1-0.0125)^{24}(0.0125)$$
$$\approx 0.0092$$

The likelihood that this first failure will occur on the 50th flight is even smaller.

$$P(R=50) = (1-0.0125)^{49}(0.0125)$$
$$\approx 0.0067$$

However, the more appropriate question involves the cumulative probability. The likelihood that the first failure occurs on or before the 25th flight can be calculated by using the principle of the complement.

$$P(R \leq 25) = 1 - P(R \geq 26)$$

For the first failure to occur after the 25th flight, the first 25 flights must all be successful. Thus:

$$P(R \leq 25) = 1 - P(R \geq 26)$$
$$= 1 - (1 - 0.0125)^{25}$$
$$\approx 0.27$$

The first catastrophic failure occurred 73 seconds after liftoff and involved the *Challenger*. It was the 25th flight in the shuttle series. Assume the estimate of one in 80 was accurate.

1. What is the expected value of the first failure?

2. Is it surprising that the first failure occurred on or before the 25th flight?

The Shuttle managers thought that 50 flights posed limited risk when the odds were one in 80.

3. What is the likelihood that the first failure would occur on or before 50 flights were completed? Should they have been more concerned?

The second catastrophe of NASA was the *Columbia* disaster which occurred on February 1, 2003 upon attempted reentry into the atmosphere. This shuttle launch was the 113th NASA shuttle and the first disaster happened during the 25th.

4. How many flights were there from the first catastrophe until the second catastrophe? Is this consistent with a 1-in-80 probability?

The original estimate was 0.0125 probability of failure.

5. If we were to use the actual shuttle experience of 135 flights, what is the probability of a failure?

6. Explain why it would be impossible to obtain a value of 0.0125 from the data for 135 flights.

In this section and the last, we developed geometric distribution. This distribution is closely linked to binomial distribution. Both distributions start with an identical process that involves repeating the same action over and over again. It could be flipping a coin, making a phone call, or launching a shuttle. The probability of success is the same for each independent repetition.

The two distributions differ with regard to the random variable of interest. With the binomial distribution, the decision maker specifies n, the number of repetitions, coin flips or phone calls. He is interested in the random variable, X, the number of successes out of n repetitions. Consequently, the maximum number of successes is n and the minimum number is 0. With the geometric distribution, the decision maker is interested in the random variable, R, the number of repetitions until the first success. The minimum number is one; there has to be at least one repetition to achieve the first success. However, the maximum is infinity. There may never be a success. Table 4.5.1 summarizes the differences.

Facet	Binomial	Geometric
Number of independent repetitions	n specified in the decision context	R random variable
Probability of success on each repetition	p	p
Number of successes	X random variable	1 stop at 1st success
Minimum value of random variable	0	1
Maximum value of random variable	n	∞
Expected value	np	$1/p$
Probability density function (pdf)	$_nC_x(p)^x(1-p)^{n-x}$	$p(1-p)^{r-1}$
TI calculator command (pdf)	binompdf(	geometpdf(
Cumulative distribution function (cdf)	no formula	$1-(1-p)^r$
TI calculator command (cdf)	binomcdf(	geometcdf(

Table 4.7.1: Comparison of binomial and geometric distributions

In the questions that follow, we demonstrate how the binomial distribution would be used to answer different types of questions about the space shuttle.

7. What is the probability of zero catastrophes in 135 flights?

8. What is the probability of exactly one catastrophe in 135 flights?

9. What is the probability of exactly two catastrophes in 135 flights?

10. What is the probability of two or more catastrophes in 135 flights?

11. Assuming a one in 80 failure rate, would it be extremely rare to have three or more catastrophic failures in 135 flights?

12. Use the above results to explain whether or not the original estimate was consistent with the actual pattern of catastrophes.

Section 4.8 Relative Frequency Approaches Theoretical Probability

The core concept of probability theory is that over the long range the observed frequency of an event will tend to approach the theoretical probability. We explore this idea by using simulated data. We compare the observed relative frequency to the theoretical values calculated with the binomial distribution function. We illustrate the concept of long-range by using different numbers of simulations. We ran the simulation 100, and 4,000 times. In addition, we repeat each simulation twice to illustrate the point that each time we obtain somewhat different relative frequencies.

The charts below illustrate how simulated data approach the theoretical probabilities for the binomial distribution when $p = 0.5$ and $n = 5$. Each chart includes two sets of simulations of one week of calls. Five calls are made each week. Each chart and table compares the results to the theoretical probabilities. A capital N is used to specify the number of simulations. A small n refers to the number of days on which calls were made.

In these examples n always equals five days of calls. In Figure 4.8.1, Figure 4.8.2, and Table 4.8.1, we use a p value of 0.5. In Figure 4.8.3, Figure 4.8.4, and Table 4.8.2, we use a p value of 0.7.

Let's review the first results for $N = 100$. When simulating five days, the number of calls answered will range from zero to five. The probability that exactly zero calls are answered is 0.0313. In the first set of 100 simulations of five days, there was only one instance of zero calls answered. The observed relative frequency was therefore 0.01. This is less than one-third the predicted value. In the second set of 100 simulations, there were three instances of calls answered. This time the observed frequency was 0.03, which is much closer to the theoretical value. For $N = 4,000$, one set of simulations recorded 120 unanswered calls and the other recorded 134 unanswered calls. The corresponding relative frequencies were 0.03 and 0.034. Clearly, with $N = 4,000$, the relative frequencies were more consistently close to the theoretical values.

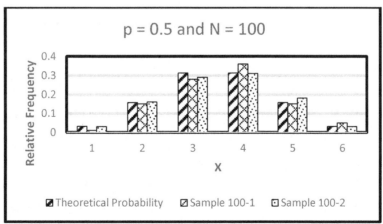

Figure 4.8.1: Binomial frequency histograms: 100 replications of simulated data ($p = 0.5$ and $n = 5$)

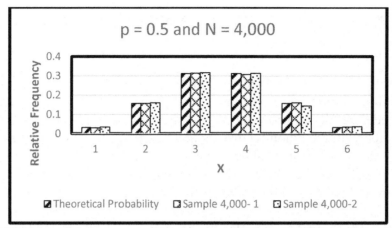

Figure 4.8.2: Binomial frequency histograms: 4000 replications of simulated data ($p = 0.5$ and $n = 5$)

		$p = 0.5$ and $n = 5$							
		$N = 100$				$N = 4,000$			
X	Theoretical Probability	Sample 100-1		Sample 100-2		Sample 4,000-1		Sample 4,000-2	
		Number	Relative Frequency	Number	Relative Frequency	Number	Relative Frequency	Number	Relative Frequency
0	0.0313	1	0.0100	3	0.0300	120	0.030	134	0.034
1	0.1563	15	0.1500	16	0.1600	625	0.156	634	0.159
2	0.3125	28	0.2800	29	0.2900	1,256	0.314	1,265	0.316
3	0.3125	36	0.3600	31	0.3100	1,228	0.307	1,253	0.313
4	0.1563	15	0.1500	18	0.1800	637	0.159	572	0.143
5	0.0313	5	0.0500	3	0.0300	134	0.034	142	0.036
	Total	100	1	100	1	4,000	1	4,000	1
	Average	2.640		2.540		2.511		2.482	

Table 4.8.1: Frequency distribution: Simulated data and binomial ($p = 0.5$ and $n = 5$)

1. What was the biggest difference between Sample 100-1 and 100-2?

2. What was the biggest difference between a sample relative frequency and the theoretical probabilities?

3. What was the smallest difference between Sample 100-1 and 100-2?

4. What was the smallest difference between a sample relative frequency and the theoretical probabilities?

5. What was the biggest difference between Sample 4,000-1 and 4,000-2?

6. What was the biggest difference between a sample relative frequency for $N = 4,000$ and the theoretical probabilities?

In this example, n is 5 and p is 0.5. Thus, the expected value for this example is 2.5.

7. What was the biggest difference between the observed average and the expected value?

8. Summarize your findings as to how well the relative frequencies match the theoretical probabilities for different values of *N*.

Another sample: *p* = 0.7

In Figure 4.8.2 and Table 4.8.2, we illustrate how simulated data approach the theoretical probabilities for the binomial distribution when $p = 0.7$ and $n = 5$. Each chart includes two simulations of five repetitions and compares the results to the theoretical probabilities. A capital N is used to specify the number of simulations while a small *n* refers to the binomial repetitions. We present the results from 100 and 4000 simulations.

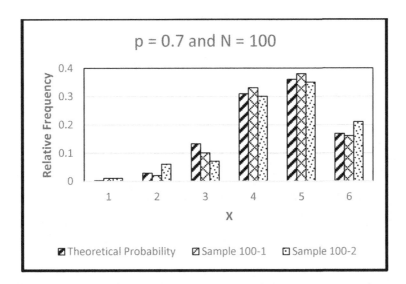

Figure 4.8.3: Binomial frequency histograms: 100 replications of simulated data
($p = 0.7$ and $n = 5$)

9. What was the biggest difference between Sample 100-1 and 100-2?

10. What was the biggest difference between a sample relative frequency and the theoretical probabilities?

11. What was the biggest difference between Sample 4,000-1 and 4,000-2?

12. What was the biggest difference between a sample relative frequency for $N = 4,000$ and the theoretical probabilities?

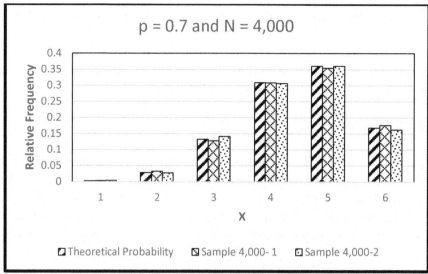

Figure 4.8.4: Binomial frequency histograms: 4,000 replications of simulated data ($p = 0.7$ and $n = 5$)

Bin	Theoretical probability	$p = 0.7$ and $n = 5$							
		$N = 100$				$N = 4,000$			
		Sample 100-1		Sample 100-2		Sample 4000 – 1		Sample 4,000 – 2	
		Number	Relative frequency	Number	Relative frequency	Number	Relative frequency	Number	Relative frequency
0	0.0024	1	0.0100	1	0.0100	11	0.003	16	0.004
1	0.0284	2	0.0200	6	0.0600	126	0.032	109	0.027
2	0.1323	10	0.1000	7	0.0700	511	0.128	563	0.141
3	0.3087	33	0.3300	30	0.3000	1,232	0.308	1,225	0.306
4	0.3602	38	0.3800	35	0.3500	1,416	0.354	1,441	0.360
5	0.1681	16	0.1600	21	0.2100	704	0.176	646	0.162
Total		100	1	100	1	500	1	500	1
Average		3.53		3.55		3.507		3.476	

Table 4.8.2: Frequency distribution: Simulated data and binomial ($p = 0.7$ and $n = 5$)

In this example, n is 5 and p is 0.7. Thus, the expected value for this example is 3.5.

13. What was the biggest difference between the observed average and the expected value?

14. Summarize your findings as to how well the relative frequencies match the theoretical probabilities for different values of N.

15. In general, for which value of p were the relative frequencies closer to the theoretical values?

Chapter 4 (Binomial and Geometric Distributions) Homework

1. Compute combinations.
    a. Find the value of $_6C_6$. Explain your answer.
    b. Find the value of $_6C_0$. Explain your answer.
    c. Find the value of $_6C_2$ - List every possible combination.
    d. Find the value of $_6C_4$ - List every possible combination.
    e. Compare and explain why the answers to c) and d) are the same.

2. A bin of external hard disks contains four disks from four different manufacturers. In how many ways can you choose two disks from the bin?

3. Compute combinations.
    a. Find the value of $_{20}C_{20}$. Explain your answer.
    b. Find the value of $_{20}C_4$. Provide a process for efficiently listing every possible combination.
    c. What combination is equivalent to part b)?
    d. Find the value of $_{20}C_{12}$.
    e. What combination is equivalent to part d)?

4. A committee of four is to be formed from a group of 25 people. How many different committees are possible?

5. In a conference of nine schools, how many intra-conference football games are played during the season if the teams all play each other exactly once?

6. A survey from Teenage Research Unlimited (Northbrook, Ill.) found that 40% of teenage consumers receive their spending money from part-time jobs. If nine teenagers are selected at random, find the probability that at least three of them will earn their spending money from part-time jobs.

7. The probability that a student is accepted to Kennedy College is 0.3. Assume five students from the same school apply. Use both the formula and the binomial pdf or the binomial cdf in your calculator to determine these probabilities.
    a. What is the probability that exactly two are accepted?
    b. What is the probability that two or fewer are accepted?
    c. The top universities in the US accept approximately one out of every 10 applicants. Calculate a) and b) again if the admission probability is 0.1.

8. Courts pick the juries randomly from specific population that is often split evenly between males and females. In many jurisdictions, civil cases have juries of six people. If a jury has six members and the population contains half males and half females, calculate the following probabilities:
    a. What is the probability of having three males and three females on the jury?

b. What is the probability of having all males?
c. What is the probability of having all jurors of the same sex?
d. What is the probability of having five or more females?

9. Now suppose, a jury has 12 members and the population have 25% minorities.
    a. What is the expected value of the number of minorities on a 12 person jury?
    b. What is the probability of not having any minorities on the jury?
    c. What is the probability of having one or fewer minority members on the jury?
    d. What is the probability of having two or fewer minorities on the jury?

10. If a student randomly guesses at 10 multiple-choice questions. Each question has four possible choices.
    a. A score of six is passing. Find the probability of getting a six or more by using the Calculator function.
    b. Given the student passed, what is the probability the student passed with the minimum grade of six?
    c. What is the expected value of the number questions the student will guess correctly?
    d. Which is more likely, two correct answers or three correct answers out of 10?
    e. A score of 9 or 10 earns an A grade. Find the probability of getting an A grade. DO this both with the formula and with binomial function for your calculator.
    f. What is the probability of getting no right answers? Compare the results to the simulation experiment in Chapter 1.

11. A marketing manager of Mm Company has monthly meetings with his five consultants to review the marketing strategies of the company. The manager will run the meeting as long as four or more of his consultants attend the meeting.
    a. If each consultant is 85% reliable in attending the meeting, what is the probability of having the meeting each month?
    b. If the manager wants to have to cancel the meetings at most 5% of the time, what standard should the manager use for the minimum number of attendees?
    c. Generate five random numbers between 1 and 100 and assign each of them to one consultant (C1 to C5). If the assigned number is 85 or less then the consultant attends; if not, the consultant misses the meeting. Record your random numbers in the first row of the following table. Then count the number of attendees and record it in the last cell of first row.

The table below will represent one year's worth of meetings

	C1	C2	C3	C4	C5	Number of attendees	Canceled Yes or No
Month 1							
Month 2							
Month 3							
Month 4							
Month 5							
Month 6							
Month 7							
Month 8							
Month 9							
Month 10							
Month 11							
Month 12							

Table 1: Simulation of consultant meeting attendance

    d. Repeat it 11 more times and record the results.
    e. Based on your result, how many times was the meeting not canceled? What is the percentage of times the meeting was cancelled?
    f. Compare the percentage with the theoretical probability.

12. In a 2007 random survey of drivers on weekend nights, the police found that the proportion of drunk drivers (0.08 alcohol level) was 0.022. The state police are planning to establish some checkpoints to catch the drunk drivers.
    a. What is the probability the first drunk driver identified occurs on the fifth pull-over?
    b. Each random stop takes approximately four minutes. The state police will stop 15 drivers on average each hour. What is the probability of catching the first drunk driver in the first hour of the check point?
    c. What is the probability that they will need to stop 30 or more drivers before finding the first drunk driver?
    d. On average how many drivers will be stopped before they find the first drunk driver?
    e. The checkpoint will be in place for four hours for a total of 60 random stops. On average how many drunk drivers will the state police find?
    f. What is the probability that they will find four or more drunk drivers during the four hours?

13. It is believed that on Saturday night between the hours of 1 am and 3 am that the percentage of drunk drivers is actually 5%.
    a. What is the probability that they will need to stop 15 or more drivers before they find the first drunk driver?
    b. On average how many drivers will be stopped before they find the first drunk driver?

c. What is the probability of not catching any drunk driver in two hours in which 30 drivers are stopped?
d. What is the probability that they will find 2 or more drunk drivers during the two hours?

14. A company wants to do a random survey of past purchasers of its product to find one who has experienced a particular problem. They know the problem exists with their product and wish to interview people to determine the impact on the customer when the problem occurs. They estimate that 1.5% of their products sold have this problem. Based on this information find the following probabilities:
    a. What is the probability of finding a first customer who experienced the problem by the 10th call?
    b. What is the probability of finding a first customer by the 50th call?
    c. What is the probability of not finding any customer with the problem in the first 50 calls?
    d. On average how many calls does an operator have to make to find a customer who experienced the problem?
    e. If one of our operators has 100 customers to call, what is the probability of finding zero customers with damaged product?
    f. What is the probability the operator will find two customers with damaged products by the 100th call?
    g. What is the probability the operator will find more than two customers with damaged products?

Chapter 4　　　　　　　　　　　　　　　　　Binomial and Geometric Distributions

Expanded Homework Questions

I. Airline Overbooking (Revisited from Chapter 1)

GLA realized that simulation of the number of airline-customer who show-up is time-consuming. They recognize that probabilistic pattern of the number who show up fits the binomial distribution. They decided to use the binomial distribution and its expected value to evaluate different overbooking strategies.

Here are a few things you need to remember from Chapter 1.
The average one-way airfare price is $300 and operating cost of the plane is $6,000 with an additional cost of $20 per customer. Flight 425 has 30-passenger seat capacity. GLA will payback $300 airfare price plus $250 compensation to the customers whose boarding is denied. Customers who cancel their flights or don't show up at the gate will pay GLA $75 out of the refund for their ticket. According to GLA flight records, on average 15% of the people who make reservation in advance don't show up at the gate.

1. No Airline Overbooking
    a. What is the expected number of arrivals if 30 customers have already reserved their tickets?
    b. What percent of the time will all 30 reserved customers appear at the gate?
    c. What is the most likely number of customers appearing at the gate? How often will this happen on average in 100 flights in which all 30 seats have been reserved?

2. Airline Overbooking
    a. If the GLA management allows 35 customers to reserve seats, how many of them will show up on average?
    b. GLA is considering an overbooking policy such that there is only a 5% chance or less that someone will have to be bumped. What is the maximum number of reservations they can accept without exceeding this 5% standard? What is the probability that someone will be bumped with this policy?
    c. GLA is considering an overbooking policy such that there is only a 10% chance or less that someone will have to be bumped. What is the maximum number of reservations they can accept without exceeding this 10% standard?
    d. Figure 1 shows the profit values for different overbooking strategies as the compensation is increased gradually from $150 to $300. Identify the optimal policy for each compensation level and on average how much money would GLA earn per overbooked flight. Which policies are never optimal? Explain why the profit curves are not strictly a monotonically increasing function of the number of reservations.

Lead Authors: Kenneth Chelst and Thomas Edwards

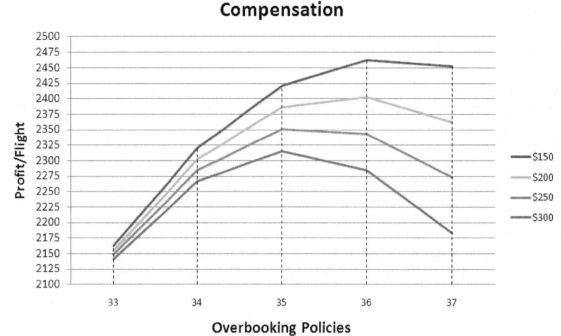

Figure 1: Optimal overbooking for different compensation values

II. Bed &Breakfast Overbooking

1. B&B Profits and Break-Even Analysis

   Niagara House is a Bed & Breakfast hotel (B&B) located in Niagara Falls, NY. It offers nine guest rooms, a full breakfast, and a private parking lot. Each room has a private bath, a queen size bed, and cable television. The rate for a room in Niagara House is $130 per night and customers have to pay a deposit for the first night in advance. If a customer wants to cancel a reservation, the customer has to cancel 24 hours in advance. There is a $25 cancellation fee that is taken out of the deposit before the remainder is refunded. From the records, seven percent of the customers cancel their reservations.

   The hotel has some daily expenses. It has a $350 fixed cost that covers the cost of the property and pays an annual income to the owners. There is a $30 variable cost per room per day that includes the cost of breakfast for two people.
   a. When does the revenue of Niagara House cover its daily expenses including its fixed cost in the case where all reservations show up?
   b. How much profit does the B&B make on a day in which it fills all the rooms with all nine reserved customers and all show up?
   c. If nine customers already reserved their rooms for the day after tomorrow, what is the probability that all rooms will be occupied?
   d. If nine customers already reserved their rooms for the day after tomorrow, what is the probability that exactly one customer will cancel his reservation 24 hours in advance? How much profit will Niagara make in this instance?

e. What is the probability that at least one customer will cancel his reservation out of nine customers?
f. What will be the profit if two of the nine reserved customers cancel their reservation and what is the likelihood of this happening?
g. What is the expected profit if they have nine reservations for the day after tomorrow?
h. If nine customers already reserved their rooms for the day after tomorrow, what is the probability that the hotel will not cover its expenses? Explain why this question does not match up with the answer provided in part a.
i. What is the probability that it will earn more than $450?

2. B&B Overbooking and Transfer

In order to cover its expenses and make profit, Niagara House should fill up as many rooms as possible. To ensure this, the hotel management uses an overbooking strategy. However, if the number of arrived customers is more than the number of available rooms, they have a standing mutual agreement to transfer them to an equivalent nearby B&B. They compensate the arrivals for their inconvenience with a gift card valued at $30.
   a. If Niagara House allows 10 customers to reserve, what is the probability that exactly one customer will be transferred to a nearby hotel and given a gift card?
   b. What will be the profit if all 10 reserved customers come to the hotel and one customer is transferred with a gift card?
   c. If Niagara House lets 10 customers reserve rooms, what is the probability that two customers will cancel their reservations?
   d. What will be the profit if two of the 10 reserved customers cancel their reservations?
   e. Would you recommend Niagara B&B adopt this overbooking policy? Explain your answer.

3. B&B Optimal Overbooking Policy
   a. What is the optimal number of reservations to accept if Niagara plans to allow for overbooking?
   b. In order to avoid losing customer goodwill, Niagara House management considers increasing the compensation value to $40. Will the previously recommended overbooking strategy still be optimal? How much will the average profit per day change? Why is it not $10?
   c. Without doing anymore detailed calculations, what do you think would happen if the compensation was increased to $50?
   d. Spring-Summer is the peak season for the Niagara House hotel. This period lasts for approximately 20 weeks. This offers an opportunity to make more profit. Because of high demand, Niagara House increases its rate for a room to $195 per night. The cancellation fee is $50 and the compensation for being transferred is increased to a $60 gift card. The daily expenses of the hotel did not change: it has a $350 fixed cost and a

$30 variable cost per room. What is the optimal overbooking strategy during the peak season?

e. What will be the expected annual profit if Niagara House follows the 9-reservation policy throughout the year? Assume that seven percent of the customers cancel their reservations and the hotel is open for 50 weeks: 30 weeks are considered off-peak and 20 weeks are considered peak season. What percent of the total profit is earned during the peak season?

f. What will be the expected annual profit if Niagara House follows the 10-reservation policy throughout the year?

g. What will be the expected annual profit if Niagara House follows the 10-reservation policy during the off-peak season and nine-reservation policy during the peak season?

h. Because of the higher cancellation fee and extremely high demand in July, the cancellation probability might decrease to only 3%. What would be the optimal policy in the month of July?

III. Service Quality Assurance

1. The manager of Koala foods wants to track each call to assess the quality of service. He is trying to figure out whether his staff needs more training or not. Therefore, AGB Company decides to select calls at random. An experienced customer service representative will listen to the conversation and categorize each as either "Satisfactory" or "Not unsatisfactory". For recent hires Mr. Smith wants AGB to ensure that by the end of a two-month training period a new operator achieves at least an 80 percent satisfactory rating. AGB picked five calls of an operator and here are the results:

Number	Quality
1	1
2	0
3	1
4	0
5	1

Table 2: Quality of service – five calls

"1" means quality is satisfying and "0" is not-satisfying.

a. Is he/she doing the job satisfactorily?
b. Can we decide about this operator based on a sample of five calls?
c. What is the likelihood that an individual who regularly achieves the standard of 80% satisfaction will be below this standard in a sample of five calls? Use this to review your answer to part b.
d. What factors will affect how large a sample of calls to listen to and evaluate?

Chapter 4   Binomial and Geometric Distributions

    e. Assume AGB will review 20 calls of an operator who generally meets the 80% standard. What is the probability that in this sample of 20, the operator will meet or exceed the standard (at least 16 satisfactory calls a day)?

    f. What is the probability that 60% or fewer of the calls will be judged satisfactory? 70% or fewer will be judged satisfactory?

    g. The manager must make a decision, as to when to send the operator to training which will cost the company more than $500. She does not want to do this unless she is reasonably sure that the low performance is not just a function of natural variability. The manager will pick a number such that the chance of seeing that many bad calls is 5% or less if the operator generally meets the 80% standard. How many bad calls should she allow out of 20? What is the corresponding percentage?

    h. Now consider the fact, AGB is going to review 50 calls of an operator. The manager will pick a number such that the chances of seeing that many bad calls is 5% or less if the operator is generally meeting the 80% standard. How many bad calls should she allow when the sample is 50? What is the corresponding percentage? Compare this answer to that of the previous problem and explain the difference.

2. Service Quality Sampling Strategy

    a. Do you think that size of our samples should depend on the cost of evaluating each call?

    b. What other factor(s) do you think should be considered when deciding on a sample size?

    c. Do you think that picking calls from only one day is a good way for sampling? Why.

    d. Do you think that the cost of sampling has any effect on the manager's decision in selecting number of the days for sampling?

    e. Do you think that categorizing the calls in just two groups is a good idea?

    f. What are the alternatives to categorize the calls?

    g. Can we use Binomial distribution to calculate the probabilities if we have more than two groups?

    h. Imagine a policy as follows for our goal of 80% satisfactory performance. If the number of bad calls is equal to or below a certain percentage X, the operator is sent to training. If it is equal to or above a certain percentage Y, the operator is certified as trained and does not need to be reviewed for another year. If the percentage is between X and Y, another sample of calls is selected and reviewed. Pick values of X and Y. Describe how you would apply this policy to an initial sample of 50 calls. Calculate the probability that an operator with a long-term 0.8 record might fall below X in a sample of 50. Calculate the probability that an operator with long-term record of 0.75 might still score above Y.

IV. Organ Harvesting

1. Organ donation and transplants save many lives that need vital organs e.g. heart, kidney, liver. Only a small percentage of deaths result in potential organ donation. In Jefferson County there are an estimated 60 deaths per year from which organs could be harvested. However, not everybody volunteers as a donor. Nationwide an estimated 25% of people sign organ donor cards. In addition, when a death occurs, hospitals try to convince family members to agree to donate organs. Some hospitals have special initiatives which have made them more successful than others in arranging family approved donations. Hospital staff efforts increase the total by anywhere from an additional 15% to 35% depending upon the hospital's effectiveness in reaching out to family members. In the County, there are currently 27 people in dire need of heart transplants within the next year.
   a. With just the pre-signed organ donor cards, what is the probability that there will be enough hearts available from 60 deaths to meet the needs of all 27 people?

   Now assume the Jefferson area hospitals have worked to reach out to family members but they are at the lower end of the range. Their efforts only add 15% to the overall total donation rate. Thus, 40% of eligible cadavers result in donated organs.
   b. On average how many hearts will become available during the year?
   c. With the pre-signed organ donor cards and 15% added by the hospital, what is the probability that there will be enough hearts available from 60 deaths to meet the needs of all 27 people?

   Now assume the Jefferson area hospitals are in the middle range and add 25% to the overall total donation rate. Thus, 50% of eligible cadavers result in donated organs.
   d. On average how many hearts will become available during the year?
   e. With the pre-signed organ donor cards and 25% added by the hospital, what is the probability that there will be enough hearts available from 60 deaths to meet the needs of all 27 people?
   f. The Jefferson area hospitals are striving to provide heart transplants to all patients on the waiting with at least 90% probability. What is the long-term average percentage of people that would have to agree to donate in order for the hospitals to have a 90% probability each year? Explain why the answer is not just 27/60 or 45%.

2. Organ Harvesting: Two neighboring counties of approximately equal size have begun coordinating their efforts. Assume there are a total of 120 eligible deaths in the two counties and that there are 54 critical patients needing heart transplants.

a. What is the long-term average percentage of people who would have to agree to donate from these two counties in order for the hospitals to have a 90% probability of providing a heart to each critical patient each year?
b. A regional consortium is being developed that involves four counties. Assume there are a total of 240 eligible deaths in the two counties and that there are 108 critical patients needing heart transplants. What is the long-term average percentage of people that would have to agree to donate from these four counties in order for the hospitals to have a 90% probability of providing a heart to each critical patient each year?
c. Discuss the impact of pooling efforts.

3. Organ Harvesting: There is an individual in Jefferson County who is extremely critical and is first on the list.
    a. On average how many eligible deaths would occur before the hospitals identify an individual with a signed organ donor card? Remember only 25% of individuals have signed cards.
    b. If the hospital is successful in its efforts to increase the rate of donation to 60% of eligible deaths, on average how many eligible deaths would occur before the hospitals obtains a heart?

Consider a situation when the donation percentage is just 25 percent. There are supplementary efforts by local hospitals. What is the probability that the first individual that has an organ donor card is associated with
    a. the first cadaver?
    b. the second cadaver?
    c. the third cadaver?
    d. the fourth cadaver or later?
    e. Assume the deaths are spread equally throughout the year with an average of five per month. What is the probability that it will take longer than a month for a cadaver from an organ donor card to become available?

Assume the hospitals' efforts increase the total percentage to 40%. What is the probability that the first individual or family to donate is associated with
    a. the first cadaver?
    b. the second cadaver?
    c. the third cadaver?
    d. the fourth cadaver or later?
    e. Assume the deaths are spread equally throughout the year with an average of five per month. What is the probability that it will take longer than a month for a cadaver from an organ donor card to become available?

Chapter 4 Summary

What have we learned?

In this chapter we have studied two related distributions, the Binomial Distribution and Geometric distribution. The formulas for these distributions replace the need for the repeated simulations we did in the previous chapter. For both distribution there are 4 common elements to each random event being analyzed:
- There are two possible outcomes.
- The p value, probability of success, is the same for each repetition.
- The outcome of each repetition is assumed to be independent.
- There are n identical repetitions of the random event.

The Binomial Distribution has two parameters: p, the probability of success on any individual trial, and n, the number of trials that are performed. The formula to find the probability distribution function for a particular number of successes, x, is $_nC_x(p)^x(1-p)^{n-x}$. We can also use the TI calculator command binompdf(n,p,x). There is no formula to find the cumulative distribution function, which adds the probabilities from 0 to x. However, we can use the TI calculator the command binomcdf(n,p,x).

The Geometric distribution has only one parameter: p, the probability of success on any individual trial. The probability that the first success occurs on the rth trial is $p(1-p)^{r-1}$. We can also use the TI calculator command geometpdf(p,r). The probability that the first success occurs in r or less days is $1-(1-p)^r$ and the TI calculator command is geometcdf(p,r).

Both these distributions can help us determine how likely a particular sequence of events would be. We can then use the results to make decisions. The three most common criteria for believing something is not random chance are: 0.1 (10%), 0.05 (5%), and 0.01 (1%).

Another important consideration is the expected value for a distribution. While the expected value is often not actually possible, it is the long run average if an event occurred many times. The best choice is the option that has the most favorable expected value.

A third consideration discussed in the chapter is the concept of conditional probabilities. A conditional probability asks the likelihood of an occurrence if we already know something that happened. The formula is $P(A|B) = \dfrac{P(A \cap B)}{P(B)}$ where A is the event we are interested in and B is the event we already know occurred.

Terms

Binomial distribution	The probability distribution that is used to find the likelihood of a particular number of successes (x) for a given number of trials (n).
$P(X = x)$	The probability of x successes: $_nC_x (p)^x (1-p)^{n-x}$
binomcdf(n,p,x)	The TI calculator command to calculate the probability of 0 to x successes out of n trials if the probability of success is p.
Conditional probability	The probability that something happens, *given* that something else happened.
Expected value	The long run average value over many repetitions.
Binomial	$E(X) = np$
Geometric	$E(X) = 1/p$
Failure	The complement of the outcome we are interested in.
Geometric distribution	The probability distribution that is used to find the likelihood of a particular trial (r) being the first success.
$P(R = r)$	The probability the rth trial will be the first success: $p(1-p)^{r-1}$
n	The number of times the random event is repeated.
$_nC_x$	The number of ways x items can be chosen out of n items. $\dfrac{n!}{(n-x)!x!}$
p	The probability that a success occurs (a number between 0 to 1).
Parameter	Values that need to be specified in order to apply a distribution.
Random variable	A variable whose possible values are numerical outcomes of a random phenomenon.
Success	The outcome we are interested in (not necessarily a good thing).

Variables and Formulas

	Binomial	Geometric
Number of independent repetitions	n specified in the decision context	R random variable
Probability of success on each repetition	p	p
Number of successes	X random variable	1 stop at 1st success
Minimum value of random variable	0	1
Maximum value of random variable	n	∞
Probability density function (pdf)	$_nC_x(p)^x(1-p)^{n-x}$	$p(1-p)^{r-1}$
Cumulative distribution function (cdf)	NA	$1-(1-p)^r$
Expected value	np	$\dfrac{1}{p}$

Chapter 4 (Binomial and Geometric Distribution) Objectives

You should be able to:

- Determine whether or not a situation has the four common elements needed to use binomial and geometric distributions.
- Calculate combination values.
- Determine whether a solution to a question requires binomial or geometric calculation.
- Calculate binomial probability distribution functions:
    - with a formula in a calculator or in Excel.
    - simulate a binomial distribution scenario.
- Calculate binomial cumulative distribution functions:
    - with a formula in a calculator.
    - in Excel.
- Calculate geometric probability distribution functions.
- Simulate a geometric distribution scenario.
- Calculate geometric cumulative distribution function.

Chapter 4 Study Guide

1. What is the binomial distribution and for what is it used? Give a specific example.

2. What are the parameters needed for the binomial distribution?

3. How do you find the expected value for the binomial distribution and what does it represent?

4. How would you simulate the binomial distribution with a value of p that is not 0.5?

5. What is the geometric distribution and for what is it used? Give a specific example.

6. What are the parameters needed for the geometric distribution?

7. How do you find the expected value for the geometric distribution and what does it represent?

8. What are the three criteria levels mentioned in the chapter?

    a. Give an example of when the highest would be more appropriate.

    b. Give an example of when the lowest would be more appropriate.

9. What does conditional probability mean?

10. If the probability of B occurring is 40% and the probability of A and B both occurring is 20%, what is the probability of A occurring given that B occurs?

CHAPTER 5:

Poisson Distribution

Section 5.0 Poisson Distribution

In the previous chapter, we developed two discrete distributions: the Binomial and the Geometric. Each of these distributions assumes that independent events with the same probability of occurring are repeated a number of times. From these assumptions, we derived general formulas for calculating individual probabilities. In addition, we presented and explained the standard expressions for the expected value. In this chapter, a new distribution is introduced: the **Poisson distribution**. Recall that the Binomial distribution has two parameters, n (the number of repetitions) and p (the probability of success). The Poisson distribution needs only one parameter: the average rate. The average rate of a Poisson distribution is usually represented by the lower case Greek letter, λ, **lambda**.

The Poisson distribution is a natural outgrowth of a random sequence of events in which the occurrence of one event has no influence on the occurrence of another random event. For example, someone falling off a bike and being rushed to a hospital has no influence as to when another person will fall off a bike and be rushed to a hospital. Each event is independent of every other event. When a random variable distributes itself according to the Poisson distribution there is a specific mathematical formula that is used to compute probabilities. This formula can determine the probability of any number of occurrences of the event, such as the number of calls to a 911 system. The entire random sequence of independent events that occur continuously is often called a Poisson Process. The number of accidents, emergencies, airplane crashes, crimes and so forth are all examples of random variables that form a Poisson Process. It is also often used to describe the random arrival of people at a box office, arrivals at a store, e-mails, or phone calls to a help desk.

This distribution is named for the 19th century French mathematician Siméon-Denis Poisson who developed it in his "Research on the Probability of Criminal and Civil Verdicts." The formula was derived as an approximation to the Binomial distribution when the number n of events is very large (i.e., in the hundreds or thousands) and p is small (i.e., less than 0.1). The derivation is beyond the scope of this text. Early in the 20th century mathematical models of telephone systems were among the first applications of the Poisson process.

Section 5.1 Deliveries in a Maternity Ward

Dr. J. D. Cox is the administrator in charge of the maternity ward at Metro Hospital, a large hospital in a metropolitan area with a population of about 4.5 million. He is interested in determining if the ward has enough staffing and space to routinely handle the daily patient volume. During the peak hours between 8a.m. and noon, the ward is staffed and designed to handle three deliveries during a four-hour period. If more than three expectant mothers arrive, Dr. Cox has to call in extra staff and make other adjustments that are less than ideal. Dr. Cox wants to determine how often the ward is unable to meet patient needs during the peak four-hour period. Based on data over the past several years, the average number of patient arrivals during this four-hour period is two.

1. If the average number of arriving women during the four-hour period is two, do you think staffing and planning for up to three deliveries is enough? Why or why not?

2. How often do you think there will be four arrivals during the same four-hour period?

3. What is the minimum number of arrivals that could occur during a four-hour period?

4. What is the maximum number of arrivals that could occur during a four-hour period?

Dr. Cox begins to think about the problem. He is particularly curious about how often there will be four or more women in labor who arrive during the four-hour period. He has experience with probability distributions, but he is unsure which distribution fits this scenario.

First, he considers the possible number of events that can occur, where an event refers to a pregnant woman in labor arriving at the ward. Next, Dr. Cox thinks about the independence of the events. That is, he wonders whether the arrival of one woman who is ready to give birth influences the arrival of any other similar patient.

5. Are the arrivals of women in labor independent of one another? Why or why not?

Dr. Cox wonders if it is possible to count the number of events.

6. Consider counting the number of events.
   a. Is it possible to count how many events have occurred? Why or why not?
   b. Is it possible to count how many events have not occurred? Why or why not?

From this line of questioning, Dr. Cox realizes that he cannot use the binomial distribution for this scenario. The binomial distribution requires an event to have two possibilities: the event occurs with probability, p, or the event does not occur with probability $1 - p$. However, Dr. Cox could not possibly know the probability of a delivery not occurring. This idea has no meaning. He realizes that the number of arrivals during a four-hour period has more than two possibilities. Therefore, he must use a different distribution.

Dr. Cox summarizes what he knows and does not know. He does some research and learns that the most appropriate distribution is the Poisson distribution.

When using the Poisson distribution, the following assumptions must be reasonable:
- The number of events can be counted in whole numbers
- Events are independent of one another
- The average rate of occurrence is known
- The number of times an event occurs can be counted while the number of times an event does not occur cannot be counted

Dr. Cox is now convinced that this distribution is appropriate for his needs because all the assumptions apply to the arrival of women in labor.

He needs to know the average number of arrivals for delivery per four-hour period (the third requirement), which is two. The Greek letter λ (lambda) is used to denote the average rate. Thus, $\lambda = 2$ arrivals per four-hour period.

Then, he can use the formula for the Poisson distribution to find the probability that they will need more staff and space during the peak four-hour period.

Formula for the Poisson Distribution

The probability of exactly x events occurring where each event previously occurred with an average rate of λ is given by the function:

$$P(X = x) = \frac{\lambda^x e^{-\lambda}}{x!}$$

where: λ = the average rate
x = number of events
e = the irrational number ≈ 2.718

Dr. Cox uses this formula to find the probability that four women in labor arrive for delivery during a four-hour period ($X = 4$ deliveries), given that the average rate is two per four-hour period ($\lambda = 2$ deliveries per four-hour period):

$$P(X = 4) = \frac{2^4 e^{-2}}{4!}$$
$$\approx 0.0902$$

Therefore, there is approximately a 9% chance that there will be four rooms in use for delivery during a four-hour period.

7. On average how many times during the year will there be four arrivals for delivery within the same four-hour period?

8. Use the formula for the Poisson distribution to find $P(X = 3)$. Interpret the meaning of this probability in terms of the problem context.

9. On average how many times during the year will there be exactly three arrivals of women in labor within the same four-hour period?

10. Use the formula for the Poisson distribution to find $P(X = 0)$. Interpret the meaning of this probability in terms of the problem context.

11. On average how many times per year will no patients arrive during a four-hour period?

Next, Dr. Cox is interested in learning the probability that the staffing and space will be sufficient to handle all of the patients. For this, he knows he should find the sum of several probabilities:

$$P(X \leq 3) = P[(X=0) \cup (X=1) \cup (X=2) \cup (X=3)]$$
$$= P(X=0) + P(X=1) + P(X=2) + P(X=3)$$

12. Find $P(X \leq 3)$. Interpret the meaning of this probability in terms of the problem context.

13. What is the complement of the event $X \leq 3$, and what does it represent in this context?

14. Based on your response in Q12, what is the probability that Dr. Cox will need to call in extra staff during a four-hour period.

15. On average, how many times per year will this happen?

16. Do you think Dr. Cox should increase the capacity of the maternity ward? Why or why not?

Formula for the Poisson Distribution with time period t

The probability of exactly x events occurring where each event previously occurred with an average rate of λ is given by the function:

$$P(X = x) = \frac{\lambda^x e^{-\lambda}}{x!}$$

where: λ = the average rate
  x = number of events
  e = the irrational number ≈ 2.718

There is another useful form of the Poisson distribution that applies to an extended period of time, t.

$$P(X = x) = \frac{(\lambda t)^x e^{-\lambda t}}{x!}$$

where: λ = the average rate per unit of time
  t = duration of time period

Summary

When using the Poisson distribution, the following assumptions must be reasonable:
- The number of events can be counted in whole numbers.
- Events are independent of one another.
- The average rate of occurrence is known.
- The number of times an event occurs can be counted while the number of times an event does not occur cannot be counted.

This list of assumptions highlights the differences between the Poisson distribution and the binomial distribution. To use the binomial distribution, one must be trying to count the number of occurrences of an event in n trials where he/she knows the probability of success on any given trial. To use the Poisson distribution, one only needs to know the average rate at which something occurs to find the probability of x occurrences in a given period of time.

Section 5.2 The CSI Team

In a suburban county outside of Detroit, Michigan, the sheriff's department only has one Crime Scene Investigation (CSI) Unit to assist the many small suburban police departments investigate major crime scenes. The county sheriff's department averages three major crime scene calls per day. These calls come in at random. Except in the case of a rare crime spree, each incident is independent of every other incident. Therefore, these calls fit the description of the Poisson Process because there is a random sequence of independent events that continuously occur. Thus, the number of crime scene calls each day can be modeled by the Poisson distribution.

The CSI team is composed of three investigators who work together to analyze major crime scenes that occur in their jurisdiction. On average, the team can investigate three crime scene calls per day, each of which averages 2.5 hours. Thus, the daily average workload matches the unit's average capability.

The members of the CSI team are always on call and are available to respond to three calls a day (i.e., 7.5 hours), five days a week. They are paid a salary based on a pay rate of $32 per hour. Even if there are not three calls on a given day, the investigators still earn a full day's pay. If the CSI team receives more than three calls on a given day, they are still required to investigate all calls in a timely manner. For more than three calls in a day, they are paid overtime at time-and-a-half their usual pay rate, or $48 per hour.

The Sheriff's Department's Chief Deputy, Dante Top, looks at how much money has been paid to the members of the CSI team over the past few years. He realizes he has a decision to make: Should he maintain the CSI team as it is now, paying for overtime when necessary? Or should he hire a new investigator and split the four investigators into two teams of two?

For this second option, a team of two investigators would take about 3.5 hours to investigate a crime scene. Each team could analyze two scenes per day. Therefore, their new salary would be based on seven hours of work and a pay rate of $32 per hour. Even if there are not two calls on a given day, the investigators still earn a full day's pay. Chief Deputy Top wants to know if it is more cost effective to maintain the current CSI team structure or to hire one more investigator.

To gain a better understanding of this problem, answer the following questions.
1. What is the total daily cost for the current CSI team if there were *three* calls in a day?
2. What is the total daily cost for the current CSI team if there were *five* calls in a day?
3. What would be the total daily cost if an additional investigator was hired and there were *three* calls in a day?
4. What would be the total daily cost if an additional investigator was hired and there were *five* calls in a day?

5.2.1 Exploring the Number of Calls in a Day

In order to decide if he should hire another investigator, Chief Deputy Top needs to find some probabilities. In particular, he wants to know the probability that the CSI team will get exactly x calls on a particular day, knowing that they average three calls per day.

5. What is the value of λ (lambda), the average number of calls per day?

6. What values can *x* be?

Chief Deputy Top thinks about using the Poisson distribution for this problem. He lists what he knows about the problem:
- He knows that the events are independent of one another.
- He knows the average number of calls per day.
- He knows the number of calls per day can be counted in whole numbers.
- He knows that there is no way to count the number of calls that do not occur in a day.

Thus, the probabilities can be modeled by the Poisson distribution.

Recall that the probability of exactly *x* events occurring, where each event previously occurred with an average rate of λ, is given by the function:

$$P(X=x) = \frac{\lambda^x e^{-\lambda}}{x!}$$

This is called the probability distribution function for the Poisson distribution. To see how this formula works, Chief Deputy Top calculates the probability of having exactly two calls to investigate in a specific day:

$$P(X=2) = \frac{3^2 e^{-3}}{2!}$$
$$\approx 0.224$$

Therefore, there is approximately a 0.224 probability that there will be exactly two crime scenes calls to investigate on any given day. This means that, on average, there will be a day with exactly two calls once every four or five days. It also means that there is 1 – 0.224 = 0.776 probability that there will either be fewer than two calls or more than two calls to investigate on any particular day.

7. Use this function to find the probability that the CSI team will have zero calls on a particular day.

Even when using a calculator, this formula is a bit cumbersome. Fortunately, modern graphing calculators have the ability to directly evaluate probabilities based on the Poisson distribution for different values of *x* and λ. Deputy Chief Top completes the following steps to find $P(X=2)$ on his calculator.

Step 1: Go to the Distributions menu and choose the command for the Poisson probability distribution function
To complete this step, Chief Deputy Top presses "2nd" and then "VARS," which gives him the Distributions (or "DISTR") menu. Then, he scrolls down until he sees "C:poissonpdf(," as shown in Figure 5.2.1.

Figure 5.2.1: The Poisson probability density function command on the graphing calculator

Step 2: Determine the average rate (λ) and the number of events (x)
The syntax for the calculator is:
$$\text{poissonpdf}(\lambda, x)$$

For example, Chief Deputy Top can find the probability that there are exactly two calls to investigate in a specific day given that the average number of calls is three. In this case, he has the following values:
  λ = the average number of calls per day = 3
  x = the number of events = 2

This calculation is shown in Figure 5.2.2.

```
poissonpdf(3,2)
       .2240418077
```

Figure 5.2.2: Finding $P(X = 2)$ on the graphing calculator

Use the poissonpdf(function on your calculator to find the following probabilities. Interpret the meaning of each probability in the context of the problem.

8. $P(X = 0)$

9. $P(X = 1)$

10. $P(X = 3)$

Chapter 5   Poisson Distribution

Complete Table 5.2.1 with the probabilities for various values of *x*. Round probabilities to four decimal places.

Calls on a Particular Day (x)	P(X = x)
0	
1	
2	
3	
4	
5	
6	
7	
8	
9	

Table 5.2.1: Probability of having *x* calls on a particular day if the average is three calls per day

11. To see a visual representation of these probabilities, create a bar graph using Figure 5.2.3. Use the probabilities from Q6. The possible numbers of calls are on the *x*-axis and the respective probabilities are on the *y*-axis.

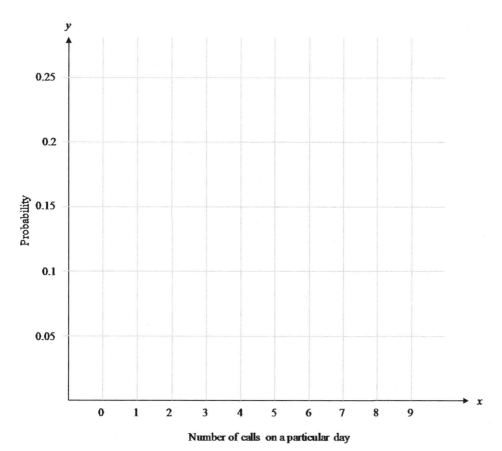

Figure 5.2.3: Create a bar graph showing the probabilities for up to nine calls on a particular day

12. What do you observe about this bar graph?

Since the CSI Unit can handle three calls per day, Chief Deputy Top is really interested in the probability that they will have more than three calls on a particular day. He starts by finding the probability that they will have *at most* three calls on a particular day. This is the same as asking for the probability of having exactly zero, one, two, or three calls on a particular day. Since it is impossible to have two different numbers of crimes scenes to investigate on the same day, Chief Deputy Top can simply add all the individual probabilities to find the probability that the CSI Unit will have at most three calls on a particular day.

13. Find the probability that the CSI team will have *at most three calls* on a particular day.

14. If Chief Deputy Top knows the probability of having at most three calls on a particular day, how can he find the probability of having more than three calls on a particular day?

15. What is the probability that the CSI Unit will have *more than three calls* on a particular day?

Next, Chief Deputy Top is curious about how often there are more than two calls on a particular day. Again, he feels that this process takes too long. Fortunately, he can use his graphing calculator to find cumulative probabilities. He completes the following steps to find $P(X \leq 2)$.

Step 1: Go to the Distributions menu and choose the command for the Poisson cumulative distribution function
To complete this step, Chief Deputy Top presses "2nd" and then "VARS", which gives him the Distributions (or "DISTR") menu. Then, he scrolls down until he sees "D:poissoncdf(," as shown in Figure 5.2.4.

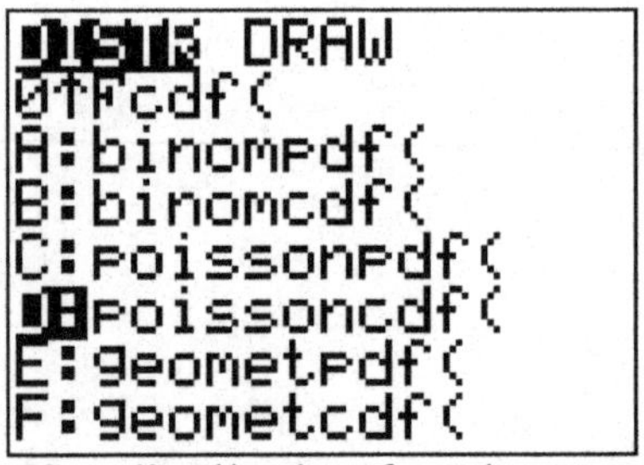

Figure 5.2.4: The Poisson cumulative distribution function command on the graphing calculator

Step 2: Determine the average rate (λ) and the maximum number of events (x)
The syntax for the calculator is:
$$\text{poissoncdf}(\lambda, x)$$

For example, Chief Deputy Top can find the probability that there are more than two calls to investigate in a specific day given that the average number of calls is three. In this case, he has the following values:
λ = the average number of calls per day = 3
x = the maximum number of events = 2

This calculation is shown in Figure 5.2.5.

```
poissoncdf(3,2)
       .4231900811
```

Figure 5.2.5: Finding $P(X \leq 2)$ on the graphing calculator

Therefore, there is approximately a 42% chance that there will be two or fewer calls on a particular day.

16. What is the probability that the CSI team will have more than two calls on a particular day?

17. Use your graphing calculator to find the probability that there will be four or fewer calls on a particular day.

18. What is the probability that the CSI team will have more than four calls on a particular day?

5.2.2 Cost of Overtime

Chief Deputy Top recently received the bi-weekly payroll statement. The cost of overtime for the CSI Unit for the two-week period seemed surprisingly large. He was puzzled since on an average day there would be no overtime. The first day of the 10-day period had four crime scene investigations. Each of the three investigators was paid for 2.5 hours at a rate of $48 per overtime hour. The total cost of overtime for that day was $(2.5)(\$48)(3) = \$360$.

Complete Table 5.2.2 to calculate the total cost of overtime for the given two weeks.

Day	Calls	Calls on Overtime	Cost of Overtime
Monday 1	4	1	1 · (2.5)($48)(3) = $360
Tuesday 1	5		
Wednesday 1	2		
Thursday 1	2		
Friday 1	1		
Monday 2	4		
Tuesday 2	2		
Wednesday 2	6		
Thursday 2	3		
Friday 2	3		
Totals	32		

Table 5.2.2: CSI calls during one two-week period

19. For how many of the 10 days was there overtime?

20. What was the total number of calls in the first week? In the second week?

21. What is the average number of calls they should have received in a week?

22. One of the weeks was below average, yet there was overtime in that week. How could that happen?

Chief Deputy Top wants to gain a clearer picture of the cost of overtime for the CSI Unit. He develops Table 5.2.3 to analyze the possible cost of overtime for one day. Then, he can use this information to approximate the expected value of overtime cost per day.

23. Complete Table 5.2.3 to calculate the cost of overtime for various values of x and the expected value of the overtime cost per day. Round probabilities to four decimal places and costs to the nearest cent.

Calls on a Particular Day (x)	Number of Calls on Overtime	$P(X = x)$	Cost of Overtime $C(X = x)$	Contribution to Expected Value of the Cost of Overtime per Day $P(X = x) \cdot C(X = x)$
0	0	0.0498	$0	(0.0498)($0) = $0
1	0			
2	0			
3	0			
4	1			
5	2			
6	3			
7	4			
8	5			
9	6			
Totals	--	0.9988	--	$239

Table 5.2.3: The cost of overtime

Using the values from Table 5.2.3, Chief Deputy Top sees that the expected cost of overtime per day is $239. On average, the cost of overtime for a bi-weekly pay period (10 work-days) would be 10 times the expected value of overtime cost per day:

Expected value of overtime cost per bi-weekly pay period = (10 work-days)($239 per day)
= $2,390

24. How does the expected cost for a bi-weekly pay period compare to the actual cost observed by Chief Deputy Top (See Q15)? Explain the difference.

Chapter 5 — Poisson Distribution

5.2.3 Exploring the Alternative Plan—Adding a Worker

Recall that the Sheriff's Department is considering adding a fourth member to the CSI team and forming two CSI teams with two investigators in each. Each investigation will then require 3.5 hours. Therefore, each team can complete two investigations per day. This increases the number of investigations the Department can do to four per day. This will clearly reduce overtime costs, but there is the increased cost of the additional investigator.

When there are three investigators, the cost per bi-weekly pay period without overtime is:
(3 investigators)($32 per hour)(7.5 hours per day)(10 days) = $7,200.

If a fourth investigator is hired at the same pay rate, the hours worked per day by each investigator will be reduced to seven.

25. Calculate the cost per bi-weekly pay period without overtime if a fourth investigator is hired. How does this compare to the present cost?

When performing analysis, Chief Deputy Top finds it helpful to start with a specific example to make sure he understands the respective costs. Thus, to compare the two plans, he first looks at the data for a particular two-week period. He then uses the probability distribution to calculate the expected value of the cost of each of the plans.

Table 5.2.4 shows the data from the same two-week period as Table 5.2.2. This time the cost of overtime is computed assuming that there were two teams of two CSI investigators. Two teams are able to cover four calls per day.

26. Complete Table 5.2.4 to calculate the total cost of overtime for the given two weeks with four investigators.

Day	Calls	Calls on Overtime	Cost of Overtime
Monday 1	4	0	0 · (3.5)($48)(2) = $0
Tuesday 1	5		
Wednesday 1	2		
Thursday 1	2		
Friday 1	1		
Monday 2	4		
Tuesday 2	2		
Wednesday 2	6		
Thursday 2	3		
Friday 2	3		
Totals	**32**		

Table 5.2.4: CSI calls during the same two-week period with two teams of two CSI investigators

Chapter 5 — Poisson Distribution

Determine which structure—the present one with one CSI team or the proposed alternative with two CSI teams—would have been more cost effective for the two-week data in Tables 5.2.2 and 5.2.4 by answering the following.

27. Calculate the total pay for the present structure (three investigators on one team) by adding the base pay (given above) to the overtime pay (see question 15).

28. Calculate the total pay for the alternative structure (four investigators on two teams) by adding the base pay (see Q22) to the overtime pay (see question 23).

29. Based on this two-week period, which structure is more cost effective?

30. Do you think this will always be the case? Why or why not?

Again, Chief Deputy Top calculates the expected value of the cost of overtime. This time, he is interested in the alternative structure of four investigators on two teams. He creates Table 5.2.5 to organize the information.

31. Complete Table 5.2.5 to calculate the alternative cost of overtime for various values of x and the expected value of the overtime cost per day. Round probabilities to four decimal places and costs to the nearest cent.

Calls on a Particular Day (x)	Number of Calls on Overtime	$P(X = x)$	Cost of Overtime $C(X = x)$	Contribution to Expected Value of the Cost of Overtime per Day $P(X = x) \cdot C(X = x)$
0	0	0.0498	$0	(0.0498)($0) = $0
1	0			
2	0			
3	0			
4	0			
5	1			
6	2			
7	3			
8	4			
9	5			
Totals	--		--	

Table 5.2.5: The cost of overtime for the alternative structure

32. What is the expected value of the difference between the current and alternative plans in overtime costs per day? Which plan has the smaller expected value of overtime cost?

33. On average, how much would be saved on overtime over a 10-day period?

34. Do you think the deputy chief should hire another CSI investigator? Why or why not?

5.2.4 Weekly Calculation of Overtime

Chief Deputy Top again reviews the overtime for the past two weeks (see Table 5.2.2). He notices that in the first week there were 14 calls, one less than the weekly average of 15. Nevertheless, there were three cases handled on overtime on Monday and Tuesday. He considers renegotiating the wage contract so that overtime is paid on a weekly basis rather than a daily basis. He is willing to increase the base salary by 10% if the investigators would accept the new overtime rules.

With the new rules, there would have been no overtime the first week. The second week, there were 18 calls, three more than the average. The investigators would have been paid overtime for three crime scene calls in the second week. The pay raise would increase the base salary to $1.1 \cdot \$32 = \$35.20$ per hour. The overtime salary would now be $1.5 \cdot \$35.20 = \$52.80$ per hour.

He decides to first see whether this policy would have saved money during the most recent two-week pay period. He will then attempt to calculate the new expected value for this policy and compare it to the current policy.

The base bi-weekly salary is currently $7,200 (calculated earlier). With the 10% pay raise, this increases by $720 dollars to $7,200 + $720 = $7,920.

The cost for overtime would be:
- Week 1 Overtime = $0
- Week 2 Overtime = (3 calls)(3 people)(2.5 hours per call)($52.80 per person per hour)
  = $1,188

Thus, the total cost would be $7,920 + $1,188 = $9,108.

The current overtime structure and salary cost is $7,200+ $2,520 = $9,720 (calculated earlier).

This would be a savings of $9,720 – $9,108 = $612 for these two weeks. The results seem encouraging. Chief Deputy Top decides to move ahead with the expected value calculation.

35. What is the cost of investigating each overtime call under this policy?

One of the fundamental theorems of the Poisson distribution is the **Addition Theorem**: The sum of Poisson random variables is also Poisson distributed.

Thus, if Chief Deputy Top adds the total number of calls in five days, it will be Poisson distributed. This makes it easier because he does not need to find each individual probability and add them together. Instead, he can combine them all together to find the overall probability.

As before, $\lambda = 3$ calls per day. Now, there is an additional variable, t, which is the number of days. In this case, $t = 5$ days. Therefore, if $\lambda = 3$ calls per day, then the five-day average will be $\lambda t = 3 \cdot 5 = 15$ calls per work-week. Now, when Chief Deputy Top applies the Poisson probability distribution function to a work-week rather than a single day, he uses λt instead of λ:

Chapter 5 — Poisson Distribution

$$P(X = x) = \frac{(\lambda t)^x e^{-\lambda t}}{x!}$$

$$= \frac{15^x e^{-15}}{x!}$$

Chief Deputy Top completes Table 5.2.6 by calculating the probability values for this Poisson distribution beginning with 16 calls. For completeness, he includes the probability that the number of crimes requiring the CSI team is 15 or less. In this case, he uses the option, poissoncdf(, on his graphing calculator. However, he notes that he must use 15 as the average now because there is an average of 15 calls per week. This is shown in Figure 5.2.6.

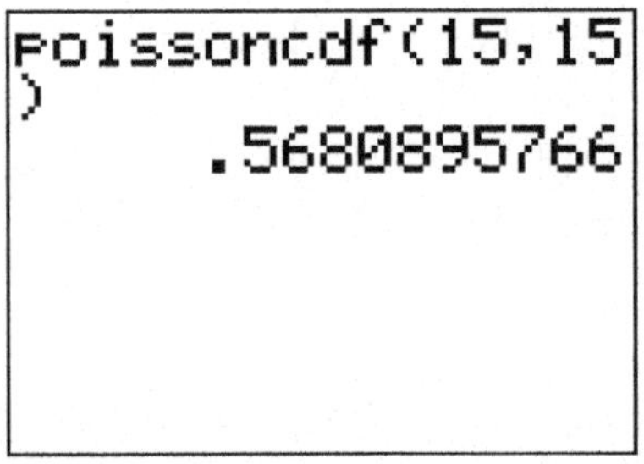

```
poissoncdf(15,15
)
            .5680895766
```

Figure 5.2.6: Finding $P(X \leq 15)$ on the graphing calculator for $\lambda t = 15$ calls per work-week

36. Why does Chief Deputy Top not need the specific probabilities for 15 or fewer calls (i.e., the first line in Table 5.2.6)?

37. Complete Table 5.2.6. Round probabilities to four decimal places and costs to the nearest cent.

Calls on a Particular Week (x)	Number of Calls on Overtime	P(X = x)	Cost of Overtime C(X = x)	Contribution to Expected Value of the Cost of Overtime per Week P(X = x) · C(X = x)
x ≤ 15	0	0.5681	$0	0.568 · $0 = $0
16	1			
17	2			
18	3			
19	4			
20	5			
21	6			
22	7			
23	8			
24	9			
25	10			
26	11			
27	12			
28	13			
Totals	--	0.9991	--	$603.92

Table 5.2.6: Cumulative Poisson probability table for $\lambda t = 15$ calls per work-week

Use Table 5.2.6 to answer the following questions.

38. What is the probability that there will be five or fewer overtime cases?

39. What is the probability that there will 10 or more overtime cases in a week?

40. What is the probability that the cost of overtime for the week will be:

   a. Less than $1,000?

   b. Between $1,000 and $2,000?

   c. More than $3,000 for a week?

The expected value of the *weekly* cost for overtime under this new policy is $603.92. Thus, the expected value of the total bi-weekly cost is:
$$\$7,920 + (2 \text{ weeks})(\$603.92 \text{ per week}) = \$7,920 + \$1,207.84$$
$$= \$9,127.84$$

Currently, the expected value of bi-weekly cost for overtime is $2,390. The expected value of the total bi-weekly cost for the current policy is:
$$\$7,200 + \$2,390 = \$9,590$$

Chief Deputy Top sees that, on average, the new policy costs less than the current policy.

41. Give an example of a situation where the new policy would cost more.

42. What is the probability that there is no overtime 10 days in a row?

Chapter 5 — Poisson Distribution

Section 5.3 Scheduled and Urgent Patients at a Health Care Clinic

Dr. Mahelia Johnson is the primary care physician at an inner city clinic in Philadelphia during the morning. This clinic provides no-cost care to residents in the community. It receives funding from private donations and grants. Dr. Johnson is on site at the clinic from 8 a.m. until noon before going to the local hospital to supervise interns and see her patients who have been admitted. With the help of nurses and physician assistants, she can see an average of four patients an hour at the clinic. This is a total of 16 patients per morning, on average.

The majority of Dr. Johnson's time is spent consulting with patients who have been scheduled a week in advance. Besides scheduled patients, Dr. Johnson wishes to reserve time for patients with an urgent problem who call up early in the day requesting an appointment that morning. On average, there are two urgent requests per morning.

1. Consider the four requirements for the Poisson distribution. Explain how each of these requirements is met in this problem scenario.
    a. The number of events can be counted in whole numbers.
    b. Events are independent of one another.
    c. The average rate of occurrence is known.
    d. The number of times an event occurs can be counted while the number of times an event *does not* occur cannot be counted.

Dr. Johnson wants to know how many time slots she should leave open each day for urgent patients. She knows that the average number of urgent patients is two per morning. However, she does not know how many patients will actually call in for urgent care. Because the requirements for the Poisson distribution are met, this actual number of urgent patients is a random variable (x) that is Poisson distributed:

$$P(X = x) = \frac{\lambda^x e^{-\lambda}}{x!}$$
$$= \frac{2^x e^{-2}}{x!}$$

2. How many urgent patient time slots do you think Dr. Johnson should leave open each morning? Explain your reasoning.

Based on the average, Dr. Johnson is thinking of scheduling only 14 patients per morning and leaving open the remaining two time slots to handle the urgent patients.

3. In what ways does using this average value not describe what might happen on any particular morning?

4. What is the probability of exactly two urgent patients in a morning?

5. If fewer than two patients call in, there will be unused slots. What is the likelihood of this happening? Interpret this probability in terms of the problem context.

There is also the possibility that there will be more urgent patients than the slots she has made available.

6. Use the answers to the above questions and the principle of complement to determine the likelihood of *more than two* urgent patients. Interpret this probability in terms of the problem context.

In an uncertain environment, there is no one policy that will turn out to be best every morning. Dr. Johnson can never be sure that the number of slots reserved for urgent patients will be needed. Nor can she ever be sure that she will have enough slots for all of the urgent patients. Dr. Johnson is torn between her desire to see as many urgent patients as possible but also not wasting any of the time slots.

Her primary concern is with the urgent patients. If she cannot treat a patient, Dr. Johnson has to send him or her to an emergency room instead. This significantly increases the cost of medical care because the clinic uses its funds to pay the hospital for the patient's care in this case. It also usually means long waits in the emergency room until the patient is seen.

She considers leaving enough open slots such that more than 90% of the time she would be able to see all of the urgent patients calling in that morning. To determine the number of slots for this goal to be met, Dr. Johnson calculates several cumulative probabilities.

7. Complete Table 5.3.1 by calculating the probabilities and cumulative probabilities for several values of x.

x	$P(X=x)$	Cumulative Probability $P(X \leq x)$
0		
1		
2		
3		
4		
5		
6		
7		
8		

Table 5.3.1: Poisson distribution with $\lambda = 2$ urgent patients per morning

8. Use Table 5.3.1 to determine how many slots Dr. Johnson should set aside to meet her 90% goal.

9. With this policy, what is the probability that there are two or more unused slots?

5.3.1 Exploring Different Options—An Economic Analysis

In order to determine the best policy, Dr. Johnson decides to perform an economic analysis. She recruits an industrial engineering undergraduate student, Marie Staats, from the local university. Ms. Staats determines that for each unused time slot, the clinic loses $138 in wasted resources. However, if an urgent patient is sent to an emergency room, the clinic must use its funds from private donors and grants to pay for the hospital's costs. The added cost for the emergency room visit is $285.

Ms. Staats calculates the expected value of the cost of various policies. She begins by finding the expected value of the cost of one reserved time slot. That is, she calculates the average cost if Dr. Johnson keeps one time slot open each morning for urgent patients.

To do so, she considers different values of x, the number of urgent patients in a morning. For each value of x, she specifies the number of unused slots as well as the number of patients sent to the emergency room.

For example, assume the policy is to reserve just one slot. Then, if there are no urgent cases, there is one unused slot. Ms. Staats uses "U 1" to represent this situation. The cost of this is $138. However, if there were four urgent patients, then three would be sent to an emergency room. Ms. Staats denotes this as "E 3". The associated cost is (3 patients)($285 per patient) = $855.

To find the expected value of the cost of this policy, Ms. Staats multiplies each cost by the respective probability and adds these values together:

$$E[C(X=x)] = \sum C(X=x) \cdot P(X=x)$$

10. Complete Table 5.3.2 to find the expected value of the cost of the policy where Dr. Johnson only has one slot reserved for urgent patients.

x	$P(X=x)$	U or E	$C(X=x)$	$P(X=x) \cdot C(X=x)$
0		U 1	1 · $138 = $138	
1				
2				
3				
4		E 3	3 · $285 = $855	
5				
6				
7				
8				
			Expected Value of the Cost	

Table 5.3.2: Expected value analysis for one reserved slot

Chapter 5   Poisson Distribution

11. Complete Table 5.3.3 to find the expected value of the cost of the policy where Dr. Johnson has two slots reserved for urgent patients.

x	$P(X = x)$	U or E	$C(X = x)$	$P(X = x) \cdot C(X = x)$
0				
1				
2				
3				
4				
5				
6				
7				
8				
			Expected Value of the Cost	

Table 5.3.3: Expected value analysis for two reserved slots

12. Complete Table 5.3.4 to find the expected value of the cost of the policy where Dr. Johnson has three slots reserved for urgent patients.

x	$P(X = x)$	U or E	$C(X = x)$	$P(X = x) \cdot C(X = x)$
0				
1				
2				
3				
4				
5				
6				
7				
8				
			Expected Value of the Cost	

Table 5.3.4: Expected value analysis for three reserved slots

13. Complete Table 5.3.5 to find the expected value of the cost of the policy where Dr. Johnson has four slots reserved for urgent patients.

x	$P(X = x)$	U or E	$C(X = x)$	$P(X = x) \cdot C(X = x)$
0				
1				
2				
3				
4				
5				
6				
7				
8				
		Expected Value of the Cost		

Table 5.3.5: Expected value analysis for four reserved slots

14. Given Dr. Johnson's concern for urgent patients and cost consciousness, what policy would you recommend and why?

After being presented the results, Dr. Johnson asks Ms. Staats to evaluate one additional policy. The doctor is considering reserving just two slots. However, if three or more urgent patients call in, she will squeeze in one more patient and come to the hospital 15 minutes late.

15. How often would she come late to the hospital?

16. Under what conditions would there be no costs associated with this new policy?

17. Complete Table 5.3.6 to find the expected value of the cost for this new policy.

x	$P(X = x)$	U or E	$C(X = x)$	$P(X = x) \cdot C(X = x)$
0				
1				
2				
3				
4				
5				
6				
7				
8				
		Expected Value of the Cost		

Table 5.3.6: Expected value analysis for two reserved slots with extra time for a third slot

18. Do you think Dr. Johnson should implement this new policy? Why or why not?

Note about Scheduled and Urgent Patients

Managing the mix of scheduled and urgent patients is even more complex than described here. All clinics have to deal with scheduled patients who do not show up for their appointments. The percentage of *no shows* can exceed 20% in some clinics. Thus, the number of scheduled patients who actually come is a random variable. That random variable follows the Binomial distribution (discussed in the previous chapter) because either a patient shows up (success) or he/she does not (failure). The total number of patients, scheduled and urgent, is then the sum of two distinct random variables. To reduce the likelihood of *no shows*, many health care providers pay staff to call all scheduled patients a day or two before their scheduled appointments. This practice can cut the rate of *no shows* by half.

Chapter 5 (Poisson Distribution) Homework Questions

1. During tax season, the city's tax department receives on average three requests per day from homeowners for a review of their property tax assessment. The number of request is random and seems to follow the Poisson distribution. Find the probability that in a randomly selected day, the number of requests is:
   a. 0
   b. 1
   c. 2
   d. 3
   e. 4 or more
   f. What is the probability that there will be two or more times the average number of requests on a random day?
   g. Provide an explanation for why you think the Poisson distribution would be appropriate?

2. Dr. Chelst teaches a large class with more than 100 students. He has found that on average five students arrive for every office hour he schedules. Because these students arrive independently of one another, he feels that random number of students who stop in follows the Poisson distribution. Find the probability that during an office hour, the number of student arrivals is:
   a. 0
   b. 1
   c. 2
   d. 3
   e. 4 or more
   f. What is the probability that there will be two or more times the average number of student visits in a random hour?

3. Dr. Chelst generally schedules two hour blocks of office hours. Dr. Chelst expects to spend on average 10 minutes with each. He is therefore concerned as to the likelihood of too many students showing up and he will have to cut his time short to accommodate the extra students.
   a. What is the probability that six students will stop by in the first hour?
   b. What is the probability that six students will stop by in the first hour and an additional six students will stop by in the second hour?
   c. What is the probability that 12 students will stop by in the two hour block? How does this answer compare to the answer in part b? Explain the difference?
   d. What is the probability that more than six students will stop by in the first hour?
   e. What is the probability more than six students will stop by in the first hour and more than six students will stop by in the second hour?
   f. What is the probability that more than 12 will stop by in the two hour block? How does this answer compare to the answer in part e?
   g. Of the above questions, which probabilities do you think are of most interest to Dr. Chelst?

Chapter 5 Poisson Distribution

4. In a one-second period of time, a radioactive source emits a random number X of particles into a counter, where X has a Poisson distribution with mean of 10 particles per second.
   a. What is the probability that no particles are actually seen?
   b. What is the probability that fewer than 10 particles are actually seen?
   c. What is the probability that there will be twice the average or even more particles in a second?

5. Vehicles pass through a junction on a busy road at an average rate of 320 per hour. The actual number in any hour is a random variable that can be described by the Poisson distribution.
   a. Find the probability that none pass in a minute.
   b. Find the probability that none pass in a five-minute period.
   c. What is the expected value of the number passing in two minutes?
   d. Find the probability that the closest integer number to this expected value actually pass through in a given two-minute period.
   e. In a two-minute period, what is the probability that there will be two or more times the average number?
   f. In a five-minute period, what is the probability that there will be two or more times the average number?
   g. In a ten-minute period, what is the probability that there will be two or more times the average number?
   h. Summarize your observations about the likelihood that more than twice the average will pass during time intervals of various lengths.

6. The coordinator of the local middle school soccer league regular gathers injury data. He found that there were a total of 72 injuries in last season's 60 games.
   a. Explain why we can use Poisson distribution with an average 1.2 injuries per game for this problem?
   b. Explain why a non-integer value is possible for the average?
   c. Find the probability there will be at most one injury in a game.
   d. Find the probability there will be one or more injuries in a game.
   e. Find the probability there will be exactly two injuries in this game.
   f. This month one team will play six games. What is the probability that in each and every game there will be at least one injury?

7. On average 6.5 traffic accidents occur on a particular stretch of road during a week. If the number of accidents follows a Poisson distribution:
   a. Find the probability that no accident will occur next week on this stretch of road.
   b. Find the probability that less than four accidents will occur next week.
   c. Find the probability that exactly five accidents will occur next week.
   d. What is the average time between accidents?

8. We continue with the previous problem in which on average 6.5 traffic accidents occur per week. We now focus on the number of accidents in a single day.
   a. What is the probability that there will be no accidents on a specific day?

b. What is the probability of having four accidents or more on a specific day?
c. On average how many days in a year will there be four or more accidents?
d. What is the probability of having six or more accidents on a specific day?
e. If six accidents occur, do you think this was simply a natural random fluctuation? What might explain what happened?

Expanded Homework Problems

I. Health Care Clinic—Reserving Slots for Random Urgent Patients

Dr. Johnson is the primary care physician at a clinic. The majority of Dr. Johnson's time is spent examining patients that have been scheduled in advance. Besides seeing patients who have scheduled appointments, Dr. Johnson must also see unscheduled patients who have an immediate urgent need to see him. Dr. Johnson wants to be able to reserve time for the urgent patients by limiting the number of scheduled patients in a day. The difficulty is deciding how many patients to schedule in advance so that he still has enough time to see any urgent patients that request time. The number of urgent patients each day is a random variable. If too many patients are scheduled in advance, then Dr. Johnson will not have enough time to see all of the urgent patients. As a result, the urgent patients will go instead to the emergency department. This is much more expensive for the patients and a poor use of emergency resources for the emergency room. However, if not enough patients are scheduled in advance, then if only a few urgent patients appear, Dr. Johnson ends up with extra time in his daily schedule. Regular patients seeking an appointment will have been delayed unnecessarily.

The data indicate that the number of urgent patients each day follows the Poisson distribution with a mean of 5 patients in a day. We explore Dr. Johnson's dilemma through a series of questions about the random pattern.

1. What is the probability of having 2 urgent patients or more in a day?

2. What is the probability of having no urgent patients in a day?

3. If the doctor keeps 4 slots available for urgent patients, what is the probability of not using all these 4 slots in a day? What is the probability that this will not be enough?

4. For this policy of 4 reserved slots, calculate the probability of 0, 1, 2, 3 and 4 unused slots. Determine the expected value of the number of unused slots. Why is this number not simply the difference between the number of slots reserved and the average number of urgent patients?

5. For this same policy, calculate the discrete probability of 0, 1, 2, … up to 10 urgent patients sent to the emergency room. Determine the expected value of the number urgent patients sent to the emergency room. Why is this number not simply the difference between the average number of urgent patients and number of slots reserved? Justify why we can stop at 10 or some other number when calculating the expected value.

6. How many slots should be reserved for urgent patients to make sure that approximately three out of every four days all of the urgent patients will be accepted?

7. With this policy, what is the probability that two or more or slots will be unused? What is the probability that two or more urgent patients will end up in the emergency room?

8. For this policy, determine the expected value of the number of unused slots. Also for this policy, determine the expected value of the number urgent patients sent to the emergency room.

9. Dr. Johnson recognizes that because of the uncertainty, there is no perfect strategy. She feels that as long as at the end of the day no more than three slots were unused, she can work during this spare time and catch up on a backlog of paperwork. She also does not want to frequently send more than one patient per day to the emergency room. Specify and analyze a policy we have not yet considered that you would recommend for reserving slots for urgent patients.

10. Can you suggest other actions that Dr. Johnson can take to deal with the dilemma she faces?

II. EMS Dispatching Problem for Poisson Distribution

1. Six neighboring cities of Oakland County in Michigan use their own dispatching centers. Assume on average each city receives on average 1.5 emergency medical calls per hour:
    a. What is the probability a city receives two calls in a random hour?
    b. What is the probability a city receives two or fewer calls in a random hour?
    c. What is the probability a city receives three calls in a random hour?
    d. What is the probability a city receives three or fewer calls in a random hour?

   The EMS units in these cities are just first responders. The patients are transported to the hospital by a private ambulance system. An EMS unit can handle on average two calls per hour. The city manager would like his city to have enough units be able to handle all of their calls in an hour 90% of the time.
    e. How many local EMS units are needed in each city to meet the 90% standard? What is the combined total of units for the six cities?

   During Friday nights, the average increases to four calls per hour.
    f. How many local EMS units are needed in each city on Friday nights to meet the 90% standard? What is the combined total of units for the six cities?

2. Now, assume that those six neighboring cities of Oakland County in Michigan use a shared EMS dispatching center. On average there are nine emergency calls per hour.
    a. What is the probability of receiving nine calls in a random hour?
    b. What is the probability of receiving nine or fewer calls in a random hour?
    c. What is the probability of receiving 15 calls in a random hour?
    d. What is the probability of receiving 15 or fewer calls in a random hour?

   Recall that an EMS unit can handle on average two calls per hour. The County Administrator would like the county to have enough units be able to handle all of the county's calls in an hour 90% of the time.

e. How many County EMS units are needed to meet the 90% standard? Compare this County total to the combined total when the six cities try to achieve the 90% standard on their own.

During Friday nights, the county average increases to 24 calls per hour.

f. How many County EMS units are needed on Friday nights to meet the 90% standard? Compare this County total to the combined total when the six cities try to achieve the 90% standard on their own.

III. Tornadoes

<Wikipedia> "A tornado is a violent, dangerous, rotating column of air which is in contact with both the surface of the earth and a cumulonimbus cloud or, in rare cases, the base of a cumulus cloud. Scientists use some techniques to rate tornadoes based on their intensity. The Fujita scale rates tornadoes by damage caused. An EF0 tornado will probably damage trees but not substantial structures; whereas an EF5 tornado can rip buildings off their foundations leaving them bare and even deform large skyscrapers."

In the United States, 80% of tornadoes are EF0 and EF1 tornadoes. The rate of occurrence drops off quickly with increasing strength—less than 1% are violent tornadoes (EF4 or stronger). Find more information about tornadoes in US and state of Oklahoma in the following website:
http://www.srh.noaa.gov/oun/tornadodata/ok/monthlyannual.php

1. The state of Oklahoma prepares for tornadoes each year and allocates enough resources to repair and fix the damage of intense tornadoes. During the peak four months from April through July, the number of intense tornados in Oklahoma each year is a random variable that follows Poisson distribution with mean seven. On average, each intense tornado costs one million dollars for the state. At the beginning of each year, the governor allocates the financial resources for handling the cost of intense tornados. If the number of intense tornados will be more than their prediction, they have to ask the Federal government for help.
   a. What is the probability of having eight or more tornados?
   b. What is the probability of having nine or more tornados?
   c. How much money should be allocated to make sure that they won't need to ask for money from the federal government with probability of 85%? How about 95%?
   d. Arkansas has a similar problem. Intense tornados during Tornado season in Arkansas follow a Poisson distribution with mean 6. How much money should be allocated to make sure that they won't ask for money from the Federal government with probability 85%? How about 95%?
   e. Now suppose that Oklahoma and Arkansas signed a collaboration agreement to have a shared budget and resources for handling the tornados. Each state has to provide 50% of the money. How much money in total should be allocated to make sure that they won't ask for money from the Federal government with probability 85%? How about 95%?

f. Compare the allocated budget of Oklahoma with collaboration and without collaboration. Are they better off collaborating?
g. Compare the allocated budget of Arkansas with collaboration and without collaboration. Are they better off collaborating?

Chapter 5 Summary

What have we learned?

In this chapter we have studied the Poisson distribution. The distribution has only one parameter, λ, the average rate of occurrence of a random event. Just as in the distributions we studied in other chapters, the occurrence of one random event does not influence the occurrence of another. For example, the occurrence of one medical emergency is independent of the next medical emergency.

The Poisson distribution is used to study the number of times a random event occurs in a time period. The probability distribution function for a particular number, x, of events is

$$P(X=x) = \frac{\lambda^x e^{-\lambda}}{x!}$$

We can also use the TI calculator command poissonpdf(λ,x). The command poissoncdf(λ,x) provides the cumulative distribution function, which adds the probabilities from 0 to x occurrences. In Excel there is one formula but within the parameters you specify whether you want the pdf or cdf.
=POISSON.DIST(x,λ,False) or alternatively =POISSON.DIST($x,\lambda,0$) refers to the pdf.
=POISSON.DIST(x,λ,True) or alternatively =POISSON.DIST($x,\lambda,1$) refers to the cdf.

Poisson distributions can also be added and the sum will still follow a Poisson distribution. Thus if λ is the average rate per day of a random event, 5λ will be the average for a five-day workweek. The more general version of the Poisson distribution uses λt instead of just λ as the mean value.

$$P(X=x) = \frac{(\lambda t)^x e^{-\lambda t}}{x!}$$

Since λ is the average rate of occurrence per unit time, the expected value for the Poisson distribution is λ for one unit in time and λt for an entire time period. As with the other distributions we have studied, the expected value for the Poisson distribution may not actually be possible.

Terms

Expected value for the Poisson distribution $E(X) = \lambda$ for one unit time and $E(X) = \lambda t$ for time period t.

λ The average rate of occurrence per unit time.

Parameter Values that need to be specified in order to apply a distribution.

Poisson distribution The probability distribution that is used to find the likelihood of a particular number of occurrences of a random event in a time period in which random events occur independent of one another.

Random event The occurrence of an event which happens at random times.

Chapter 5 (Poisson) Objectives

You should be able to:

- Determine whether or not the Poisson distribution would be appropriate for a situation.

- Determine when to use λ and when to use λt as the Poisson parameter.

- Determine whether the probability distribution function or cumulative distribution function is the appropriate function to use in a specific problem context.

- Calculate Poisson probability distribution functions with a TI calculator and Excel.

- Calculate Poisson cumulative distribution functions TI calculator and Excel.

Chapter 5 Study Guide

1. Provide a specific example when you would use the Poisson distribution to model a random pattern.

2. Provide a specific example when you would use λ and when to use λt as the Poisson parameter.

3. Explain one difference between the Binomial distribution and the Poisson distribution.

4. For λ equal to seven calls per hour, determine the probability that there will be exactly five calls in one hour.

5. For λ equal to seven calls per hour, determine the probability that there will be five of fewer calls in one hour.

6. For λ equal to seven calls per hour, determine the probability that there will be 20 or fewer calls in a three hour period.

CHAPTER 6:

Normal Distribution

Section 6.0 The Normal Distribution

The Normal distribution is the most widely recognized of all probability distributions. It is a continuous distribution, which means its graph has no gaps. The shape of its graph is classically described as bell-shaped. It is symmetric around the **mean** with long tails. It was originally discovered in the early part of the 18th century by the French mathematician, Abraham De Moivre and independently rediscovered in the 19th century by the German mathematician, Carl Gauss. The distribution is described by a complex formula that cannot be used analytically to answer probability questions. Instead, all probability questions are answered by inserting values into a calculator or spreadsheet.

The normal distribution is the most important probability distribution that is used in statistics. This is due to a number of desirable mathematical properties that normal distributions have. Statisticians have been able to solve many problems in statistics only by assuming a normal distribution and using its mathematical properties. This is because there are many variables, such as intelligence or height, that when plotted appear similar to it.

However, the normal distribution is only a model. There are, in fact, no variables that completely match it. Furthermore, there are many other models, such as exponential growth (e.g., bacterial populations) or exponential decay (e.g., carbon dating). Each possibly has a large number of important variables that when plotted appear similar to them as well. The existence of other models has a major impact on the consequences of assuming the normal distribution if it is not reasonable to do so.

The normal distribution arises naturally when the random variable is the sum or mean of several other independent random variables. The distribution has two independent parameters: the **mean** and the **standard deviation**. Sometimes the variance is used instead of the standard deviation. An important concept is that any normal distribution with a mean of μ and a standard deviation of σ can be transformed into an equivalent normal distribution with a mean of 0 and a standard deviation of 1. This new distribution is called the **standard normal** distribution. It can be used to quickly estimate with good accuracy probability questions about any normal distribution, sample mean, or sample proportion. With the advent of technology, calculators or computers with appropriate software are used to provide answers to probability questions based on normal distributions. For example, a graphing calculator determines the probability by finding the corresponding area under the graph of a normal distribution function. In this chapter, we use the graphing calculator as the primary tool to calculate probabilities.

Section 6.1 Cutting Fabric for Parachutes

Skydiving equipment has advanced considerably over the last several years. Round parachutes are seldom seen these days. They have been replaced by modern, rectangular "ram-air" canopies that have better directional control and offer softer landings. Ram-air canopies are made of a series of inflatable tubes or "cells," connected side-by-side along their length. Each cell is designed to form the cross section of an airfoil. When the parachute inflates, it forms a wing-shaped canopy, ready for flight. The front of each cell is open to the air, and the back is sewn closed. Once inflated, the ram-air canopy is a semi-rigid, rectangular plane, similar to an airplane wing. Figure 6.1.1 shows a ram-air canopy parachute with seven cells. (from http://www.uspa.org/AboutSkydiving/Equipment/tabid/128/Default.aspx)

Figure 6.1.1: A ram-air parachute
(from http://ozcrw.tripod.com/crw_canopies.htm)

The fabric most often used to make parachutes is called rip stop nylon. It is commercially available in 20-meter rolls that are 150 centimeters wide. Depending on the final dimensions of the parachute, the size of each cell will vary as will the length of fabric that must be cut. Regardless of the size, the fabric must be cut with great precision. If the pieces are cut too small, the material cannot be used. If they are cut too large, the material can be used, but you will have wasted some of your fabric. You may need to re-cut the fabric down to size. The margin of error for cutting the fabric is five millimeters.

1. What is the relationship between meters and centimeters? ... between centimeters and millimeters?

6.1.1 Sampling the Cut Pieces

Production rate is a measure of productivity. The production rate represents the output after a given amount of time. Sky's the Limit is a small company that makes parachutes. Twin brothers Tom and Ken Speedy work in the cutting department at Sky's the Limit. They cut the large panels of fabric used in making the parachutes. In this case, their production rates would be the number of pieces each cuts per day. Tom and Ken Speedy are cutting pieces that are supposed to be 50 cm wide and two meters long. On a particular day, the company decides to take random samples of 10 pieces of Tom's and Ken's work. The company will measure each piece to see if it

Chapter 6   The Normal Distribution

is within specifications: 2 m long ± 5 mm and 50 cm wide ± 5 mm. The length and width of each piece of work that was sampled appear in Table 6.1.1. Ray Monitor, the production manager at Sky's the Limit, decides he needs to compute some statistics to help him make sense of the data. He decides to compare the *means* of the lengths of the two samples.

	Tom		Ken	
Sample	Length (m)	Width (cm)	Length (m)	Width (cm)
1	2.004	50.3	1.999	49.9
2	1.999	50.3	2.001	50.1
3	2.002	50.2	1.999	50.3
4	1.997	50.3	2.000	49.9
5	2.003	49.7	1.999	50.0
6	2.000	49.8	2.002	49.7
7	2.000	49.7	1.999	50.1
8	1.998	50.2	1.999	49.9
9	1.996	49.8	2.002	50.1
10	2.001	49.7	2.000	50.0

Table 6.1.1: Lengths and widths of two samples measured to the nearest millimeter

> The *mean* of a sample is just the arithmetic average:
> Add each of the sampled items and divide by the number of items.

> If the elements of a data set are represented by x_i, the usual notation for the mean of the data set is $\bar{x}$.

2. Compute the mean of the lengths of each worker's sample. What do you observe?

The mean is one way to describe the center of a set of data. Ray was surprised that the two means were equal. When he looked at the two sets of data, they somehow looked different to him.

3. Do the two sets of lengths look different to you? If so, how are they different? If not, other than having the same mean, what similarities did you see?

Then Ray had an "Ah-ha" moment! He noticed that Tom's lengths varied between 1.996 and 2.004; while Ken's varied between 1.999 and 2.002. Ray can use a statistic called the *range* to express what he noticed mathematically. He just subtracted the smallest number from the largest number in each data set. The range is one way to measure how spread-out the individual data points are.

4. Compute the range of the lengths for each worker's sample. What do you observe?

Now Ray wonders if he will observe the same things about the width data.

5. Compute the mean and the range for each worker's sample of widths. What do you observe?

Ray finds the width data difficult to make sense of. Although the mean *and* range for each worker's width data are the same, he is still sure that he can see a difference.

6. Do you see any difference in the two width data sets? If yes, how are they different? If not, besides having the same mean and range, what similarities do they have?

Ray decides that he needs another statistic to measure the spread of the data. He decides to use a statistic called the ***standard deviation***. The standard deviation uses the difference, or deviation, of each data point from the mean to measure the spread of the data. So first, for each data point, x_i, the difference $\bar{x} - x_i$ is computed. Next, finding an average difference would make sense. However, some of the differences are going to be positive, and some negative. If they were just added together, the positives and negatives would cancel each other. To avoid that, each of the differences is squared. Then the squared differences are added. If there are n elements in the data set, this can be written as:

$$\left(\bar{x} - x_1\right)^2 + \left(\bar{x} - x_2\right)^2 + \ldots + \left(\bar{x} - x_n\right)^2 = \sum_{i=1}^{n}\left(\bar{x} - x_i\right)^2$$

7. Why does adding the squared differences avoid the problem of positives and negatives canceling each other?

To complete the process of finding a standard deviation, first we divide by $n-1$. (The reason for dividing by $(n-1)$ rather than n is technical and beyond the scope of this text. An explanation can be found in many statistics books.) Then we take the square root of the result, so that our units of measurement will be the same. The symbol for the standard deviation is S_x and we can write:

$$S_x = \sqrt{\frac{\sum_{i=1}^{n}\left(\bar{x} - x_i\right)^2}{n-1}}$$

A note on notation: As you have already seen, we use two different symbols to represent the ***mean*** of a set of data: $\bar{x}$ and μ. We also use two different symbols for the ***standard deviation*** of a set of data: S_x and σ. As you will soon see, we also use two different symbols for the ***variance*** of a set of data: S_x^2 and σ^2. You will need to be flexible in interpreting and using these symbols in this text as well as with your calculator. The first symbol in each pair is typically used to characterize a sample while the second generally applies to a population.

8. Using this process, find the standard deviation for each of the worker's width data. What do you observe about the two standard deviations? Why do you think that happened?

Sometimes it is useful to work with a statistic called the ***variance***. The variance is simply the square of the standard deviation. Therefore, it is denoted by S_x^2. Notice that the square root of the variance is the standard deviation, which is as it should be:

$$\boxed{\sqrt{S_x^2} = S_x}$$

9. Calculate the variances of Tom's and Ken's samples.

Notice also that the variance is always computed as part of the computation of the standard deviation. Why, then, do we have two different statistics? The first reason is that the units of measure for the variance might not be meaningful. Even when they are meaningful, they are not appropriate to the problem context. For example, in our width problem above, the units in which the widths were measured are centimeters. However, the variance squares differences that are measured in centimeters; the units of the variance are *centimeters squared*, which is not a unit of distance or area. It is inappropriate to use a unit that is not a unit of distance to measure variation in a distance. Thus, when we take the square root of the variance to obtain the standard deviation, we are also converting the units back to centimeters.

6.1.2 Putting Two Pieces Together

There is a second reason that we use two different statistics, the variance and the standard deviation. The standard deviation, which is the more meaningful statistic, lacks an important property that the less meaningful statistic, the variance, has. To see this, consider a third worker who takes a panel cut by Tom and another cut by Ken and stitches them together. Before he does so, he overlaps the panels by one centimeter. On average, how wide would the resulting panel be? How much variability should we expect?

The mean width of the samples taken from both Tom's and Ken's work is 50 cm. On average the width of two panels sewn together with a one cm overlap will be 99 cm. Now, how much variability will there be, on average?

The standard deviation and the variance are both measures of variability. Perhaps it seems that we could just add the two standard deviations or the two variances together, just like we did for the mean. Let's see what happens when we do.

Sample	Tom's Width (cm)	Ken's Width (cm)	Width of doubled panel (cm)
1	50.3	49.9	50.3 + 49.9 − 1 = 99.2
2	50.3	50.1	99.4
3	50.2	50.3	99.5
4	50.3	49.9	99.2
5	49.7	50.0	98.7
6	49.8	49.7	98.5
7	49.7	50.1	98.8
8	50.2	49.9	99.1
9	49.8	50.1	98.9
10	49.7	50.0	98.7

Table 6.1.2: Widths of double-panels made from two workers' samples

Table 6.1.2 shows the widths of a double panel when one of the panels from Tom's sample and one of the panels from Ken's sample are overlapped and sewn together. You can use a graphing calculator to compute the mean, standard deviation, and variance of the widths of the doubled panels. To begin, in the mode screen, choose to display two decimal places. Then use STAT-EDIT to access the data lists and enter the "Width of doubled panel" data into List 1, as shown in Figure 6.1.2. Next, use STAT-CALC to access the 1-Var Stats feature to calculate the one-variable statistics, as shown in Figure 6.1.3. The calculator displays only the mean, $\bar{x}$, and the standard deviation, S_x. From the screen capture, we can see that $\bar{x} = 99$ cm and $S_x = 0.33$ cm. We must use S_x to calculate the variance, S_x^2. The screens in Figure 6.1.4 show how to access S_x from the VARS-Statistics menu, and in the last screen, we see that $S_x^2 = 0.11$ cm^2.

Figure 6.1.2: Entering the data into a calculator

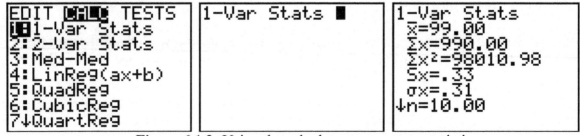

Figure 6.1.3: Using the calculator to compute statistics

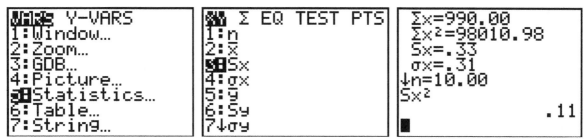

Figure 6.1.4: Using the standard deviation to find the variance

1. Add the means of Ken's and Tom's samples that you calculated earlier and subtract the one cm overlap. Compare your answer to the mean of the doubled panels. What do you notice?

2. Add the variances of Ken's and Tom's samples that you calculated earlier. Compare your answer to the variance of the doubled panels. What do you notice?

3. Add the standard deviations of Ken's and Tom's samples that you calculated earlier and compare your sum to the standard deviation of the doubled panels. What do you notice?

You should have found that when the elements from two data sets are added, the means and variances can also be added to find the mean and variance of the new data set. However, the standard deviations cannot be added to find the standard deviation of the new data set.

Why can't the standard deviations be added? You might recall that the square root of a sum is not the same as the sum of the square roots. For example, $\sqrt{4+9} \neq \sqrt{9}+\sqrt{4}$, because

$$\sqrt{4+9} \neq \sqrt{9}+\sqrt{4}$$
$$\sqrt{13} \neq 3+2$$
$$3.61... \neq 5$$

So, when two variances are added, you get the variance of the data set formed by adding the two sets of data. However, the standard deviation is the square root of the variance. The square root of a sum of two variances does not equal the sum of the square root of each variance.

> The sum, T, of two independent normally distributed random variables, X and Y, is also normally distributed.
> - The **mean** of the new random variable is the sum of the means:
> $$\mu_T = \mu_X + \mu_Y.$$
> - The **variance** of the new random variable is the sum of the variances:
> $$\sigma_T^2 = \sigma_X^2 + \sigma_Y^2.$$
> - The **standard deviation** of the new random variable is not the sum of the standard deviations:
> $$\sigma_T = \sqrt{\sigma_X^2 + \sigma_Y^2}$$

6.1.3 Putting the Whole Canopy Together

Recall that the mean and range of Tom's and Ken's samples of parachute panels were the same, but that Ken's sample had the smaller standard deviation. From past experience, Ray Monitor, their supervisor, knows that when something is cut to a specification, the cuts are approximately normally distributed about the mean.

To see what that means, let's graph a normal distribution based on Tom's width data. We cannot know the mean, μ, nor the standard deviation, σ, of the widths of all of the panels Tom has ever cut. However, we do have the mean, $\bar{x}$, and the standard deviation, S_x, of the sample that he cut for Ray Monitor. We can use those statistics, $\bar{x}$ and S_x, as estimates of the actual parameters, μ and σ, of Tom's widths. Figure 6.1.5 shows the process of graphing a normal distribution of widths that have a mean of 50 cm and a standard deviation of 0.28 cm. Notice the bell shape. The vertical line in the fourth graph goes through $x = 50$ cm, which is the mean of the distribution.

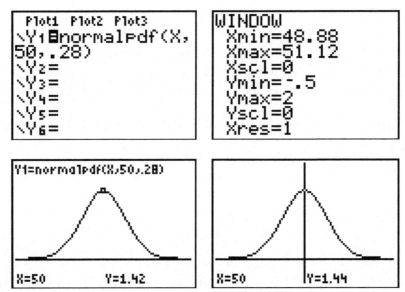

Figure 6.1.5: Graph of a normal distribution with $\mu = 50$ and $\sigma = 0.28$

1. What proportion of the area under the bell-shaped curve and above the x-axis appears to be to the left of the vertical line? What proportion appears to be to the right?

2. What do you think is the probability that a value of x is greater than 50? ... less than 50?

3. Use the mean and standard deviation of Ken's width data to generate a similar graph. What proportion of the area of your graph appears to be on either side of a vertical line through the mean?

4. What do you think is the probability that a value of x on your graph is greater than 50 cm or less than 50 cm?

In order to use a normal distribution, Ray Monitor needs to know the mean and standard deviation for the widths of all of the panels that *each* of his workers cuts, not just Ken and Tom. That information would not be available unless every panel had been measured after it was cut. That would have been very costly. Instead, Ray decides to use the mean and standard deviations from the samples that were measured. He reasons that these statistics are probably good estimates of the actual mean and standard deviation. Recall that the mean of each of their samples was 50 cm, but the standard deviation of Tom's sample was 0.28 cm, while the standard deviation of Ken's was 0.16 cm.

When data are normally distributed with mean, μ, and standard deviation, σ, the distribution features of a calculator can be used to find the probability that a particular value falls within a given range. For example, the range of interest in this problem is widths between 49.5 cm and 50.5 cm, because widths in this range are within specifications. So, a key question that Ray asks himself is, "What is the probability that the panels Tom cuts meet the width specifications?" In other words, what is the probability that Tom cuts a panel with a width between 49.5 cm and 50.5 cm?

We can answer this question using the ***normal cumulative distribution function*** (cdf), as shown in Figure 6.1.6. First we use the mode key to set the calculator to display 3 decimal places. The key parameters are the mean (50 cm), standard deviation (0.28 cm), and the endpoints of the interval, 49.5 cm and 50.5 cm. The syntax is normalcdf(49.5,50.5,50,0.28). From the Figure 6.1.6, we can see that the probability Tom cuts a panel with its width within specifications is 0.926, or 92.6%.

Figure 6.1.6: Setting up a normal cdf on a graphing calculator

Table 6.1.2 presents a sample of 10 measurements from Tom's work.

5. What percentage of Tom's sample data meets the specifications? Is this consistent with what the normal distribution predicts?

6. Using the mean and standard deviation of Ken's sample, find the probability that Ken cuts a panel with the width within specifications.

Ray Monitor knows that Ken and Tom are two of his best workers. He decides to collect a random sample of 25 cut panels from each of his 50 workers. When he computes the statistics from the large sample, he finds that the mean is 50.05 cm and the standard deviation is 0.3 cm

7. Assume a normal distribution and use the mean and standard deviation of the large sample. Find the probability that the width of a panel cut by any one of Ray's workers meets specifications.

8. In a production run of 7,000 panels, on average, how many will meet the width specifications? How many will not?

Each panel cut to the correct width specifications is worth $6.00. Each panel that does not meet the width specifications can be used for other products, but is then worth only $1.50.

9. Find the average value of a production run of 7,000 panels.

Seven panels are stitched together to form a parachute canopy. Remember that the panels are overlapped 1 cm for each seam. There will be six seams in a rectangular parachute as shown in Figure 6.1.1. The seven panels are chosen randomly from all of the panels cut by all of the workers. Now, suppose that the specifications for the entire parachute canopy indicate that the width should be 344 cm ± 2 cm. Ray Monitor would now like to know the probability that a completed parachute canopy meets this width specification.

First, he has to find the mean and the standard deviation of the width of an entire parachute canopy after it has been sewn together. Recall that the mean of a sum is the same as the sum of the means. He reasons that, on average, the width of a finished canopy will be 7 × 50.05 cm less the total of 6 cm of overlap. So, the average width is going to be 344.35 cm.

> The sum, T, of n independent identically distributed random variables, $T = X_1 + X_2 + \ldots X_n$, tends towards a normally distributed random variable.
>
> Protocols assume that if $n \geq 5$ random variables, the normal distribution is a reasonable approximation to the distribution of the sum of n random variables. This is true even if the individual random variables are not normally distributed.
>
> - The mean of the sum of n random variable is simply n times the individual value.
> $$\mu_T = n \cdot \mu_X$$
> - The variance of the new random variable is the sum of the n identical variances.
> $$\sigma_T^2 = n \cdot \sigma_X^2$$
> - The standard deviation of the new random variable
> $$\sigma_T = \sqrt{n \cdot \sigma_X^2} = \sigma_X \sqrt{n}$$

Next, he needs to calculate the standard deviation for the width of the seven panels that are sewn together to form the canopy. However, he remembers that he cannot just add the standard deviations, because the square root of a sum is not the same as the sum of the square roots. He must square the standard deviation to find the variance, add the variances together, and then take the square root. The standard deviation of the large sample was 0.3 cm and the variance is 0.09 cm². The total variance of the seven pieces is seven times the individual variance.

$$\sigma_{7X}^2 = (7)(0.09 \text{ cm}^2) = 0.63 \text{ cm}^2$$

$$\sigma_{7X} = \sqrt{0.63 \text{ cm}^2} \approx 0.794 \text{ cm}$$

So, the standard deviation for the width of seven panels sewn together is 0.794. Now Ray is ready the use the normalcdf(function to compute the probability that a finished parachute canopy will meet the width specifications. Figure 6.1.7 displays Ray's result. Almost 98% of the parachutes will meet the width specifications.

```
normalcdf(344-2.
0,344+2.0,344.35
,.794)
          .9796095557
```

Figure 6.1.7: Computation of the probability of meeting the width specification

10. Why were (344 – 2.0) and (344 + 2.0) used in setting up the normal cdf?

Excel equivalent to normal cdf

Excel has a function that calculates the cumulative distribution for the normal distribution. Its input parameters are the x value of interest, the mean and standard deviation. It requires the user to specify that he wants the cdf by typing *TRUE* in the last position in the function.

=NORM.DIST(x, mean, standard_dev,cumulative)

For example, if we wanted to know the probability that a canopy will be less than 49.5 cm

$P(X \leq 49.5)$ =NORM.DIST(49.5, 50, 6,True)

However, if we wanted to know the probability that a canopy will be more than 50.5 cm, we need to use the concept of the complement.

$P(X \geq 50.5) = 1 - P(X \leq 50.5) = 1 -$ NORM.DIST(50.5, 50, 6,True)

In Figure 6.1.7 we were interested in the probability of falling within the range of 342 cm and 346 cm. In terms of the graphing calculator, this is equivalent to finding the area under the curve between 342 cm and 346 cm. However, this can also be found by finding the area to the left of

346 cm and subtracting from it the area to the left of 342 cm. Excel only generates the cumulative distribution. Thus, to find a probability for the range, it is necessary to subtract cumulative probability for the smaller value from the cumulative probability for the larger value.

$$P(342 \leq X \leq 346) = P(X \leq 346) - P(X \leq 342)$$

This is represented in Excel with the following expression.

= NORM.DIST(346, 344.35, 0.794,True) – NORM.DIST(342, 344.35, 0.794,True)

Which results in the following values.

= 0.9811 – 0.0015 = 0.9796

6.1.4 Waste vs. Production–An Economic Analysis

The two brothers, Tom and Ken Speedy are both up for raises. Unfortunately, their supervisor, Ray Monitor, knows that the company can only afford a small raise for one of them. Ray's dilemma is deciding which brother is more deserving of the raise.

Ray knows that on an average day, Tom produces 180 cut panels, while Ken produces 171. Ray also remembers that although the mean widths of their production samples were equal, Ken's sample had a smaller standard deviation. Ray decides to see if this matters. Recall from the previous section that 92.6% of Tom's cut panels met specifications, while 99.8% of Ken's did.

Ray decides to calculate the value per day of each worker to the company. Tom produces more cut panels per day, but only 92.6% of them are usable for parachute canopies. The other 7.4% are unusable for parachute canopies and must be used for other products. Ray also knows that the value to the company of each usable panel is $6.00, while the value to the company of an unusable panel is $1.50. Ray computes Tom's per day value to the company this way:

(0.926)(180)($6.00) + (0.074)(180)($1.50) = $1,020.06.

1. Explain each part of the formula above that Ray used to compute Tom's per day value to the company.

2. Using a similar formula, compute Ken's per day value to the company.

3. Based on the per day value to the company of these two workers, which one should get the raise?

Section 6.2 Automobile Battery Warranties

Alex Murphy is a senior at Chesterfield High School in Chesterfield Township, Michigan. In a case of what he likes to call Murphy's Law, the battery in his recently purchased used car had just died. He needed to replace it as soon as possible. He intended to keep this car through his senior year of high school as well as through his four years of college, so a reliable car battery with a high-quality warranty was essential. Alex decided to purchase a car battery that costs $110 and includes a 36-month full replacement warranty with an additional 6-month pro-rated warranty. Let T represent the time, in months, from the purchase of the car battery until it fails. Table 6.2.1 shows the replacement values for the warranty period.

Time from battery purchase to failure (months)	Replacement value (% of original purchase price)
$T \leq 36$	100
$36 < T \leq 39$	75
$39 < T \leq 42$	50
$T > 42$	0

Table 6.2.1: Replacement values for battery failure within warranty period

6.2.1 Using a Normal Distribution to Find the Probability of Battery Failure

After doing some additional research, Alex found an old article in an automotive magazine with some data about battery life for the same battery that he was considering purchasing. The magazine collected data on a sample of 50 such batteries. Alex put the data into his graphing calculator and used it to compute some statistics. He also used his calculator to draw a histogram of the data. The set-up of Alex's calculator appears in Figure 6.2.1, and his results are displayed in Figure 6.2.2.

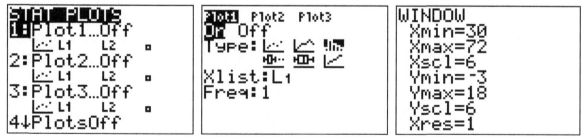

Figure 6.2.1: Alex's calculator set-up

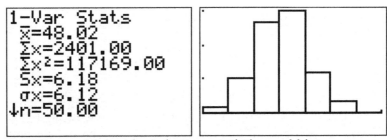

Figure 6.2.2: Alex's battery statistics and histogram

Alex thought he remembered seeing histograms with the same shape as his battery histogram when he learned about normal distributions in his math class. Looking at the histogram and the statistics, Alex noticed that the mean battery life was a little more than 48 months and the standard deviation was close to six months. He chose to assume that the battery life is normally distributed with a mean of 48 months and a standard deviation of six months. Alex decided to look at the graph of the ***normal probability density function*** (pdf) and consider whether it seemed to be a good representation of the data in his histogram. He did this on his graphing calculator by (a) opening the Y= menu, (b) pressing 2nd VARS to view the distributions menu and selecting normalpdf(, and finally (c) specifying the variable, the mean, and the standard deviation. Figure 6.2.3 shows the calculator screens for these steps.

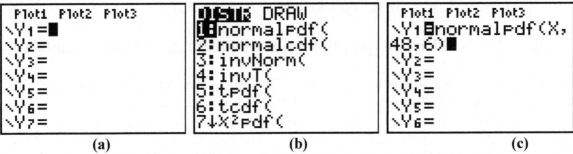

Figure 6.2.3: Steps to graphing a normal distribution on a graphing calculator

The most challenging step in graphing the normal distribution is setting up an appropriate viewing window. In order to set the Xmin and Xmax, Alex reasoned that the mean should be in the middle, half-way between the Xmin and Xmax,. He also thought that it made sense to scale the *x*-axis using the standard deviation; which is six. His real challenge came in setting-up the *y*-axis. At first, he couldn't see anything, but then he remembered learning that the area between the graph of a normal distribution and the *x*-axis is always one. Since his *x*-values were fairly large, he decided to use fairly small values for *y*. The viewing window Alex used and his graph of this normal distribution are given in Figure 6.2.4.

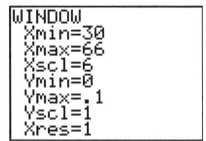 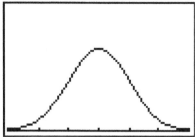

Figure 6.2.4: Alex's viewing window and graph of the normalpdf with $\mu = 48$ and $\sigma = 6$

1. What are some features that you notice about the graph of this normal distribution?

Alex now wanted to find the probability that the battery he chose will fail within the ranges given in the terms of the battery warranty. At first he was not concerned about the probability of the battery failing within the first 36 months. In that case he would get 100% of his money back. Instead, he considered the first range in the pro-rated warranty period. If the car battery fails between 36 and 39 months of his purchase, Alex would be refunded 75% of the original purchase price.

2. How much money would Alex get back if the battery fails between 36 and 39 months?

One way to find the probability is to use the normalcdf(feature on the graphing calculator. Recall that for this particular car battery, Alex assumed that $\mu = 48$ months and $\sigma =$ six months. He wanted to know the probability that the battery would fail between 36 and 39 months of purchase. Figure 6.2.5 shows the calculation. The output tells us the probability that the battery would fail between 36 and 39 months of the purchase is approximately 4.4%.

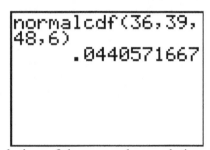

Figure 6.2.5: Calculation of the normal cumulative distribution function

3. On average, out of every 100 of these batteries that are made, how many fail between 36 and 39 months after purchase? How many out of every 1,000?

Alex likes to visualize things whenever possible. He remembered a second way to use his calculator to find this probability. This time, he used the graph he created earlier and option 7 under the CALC menu, which looks like this: $\int f(x)dx$. This represents a concept from calculus called the definite integral. To use this feature of the calculator, while looking at the graph, Alex

    (a) pressed 2nd TRACE to call up the CALC (calculate) menu and pressed number 7,
    (b) entered the lower limit, 36, for the range of months he was interested in, then

(c) entered the upper limit, 39. Figure 6.2.6 shows Alex's calculator screens for these steps.

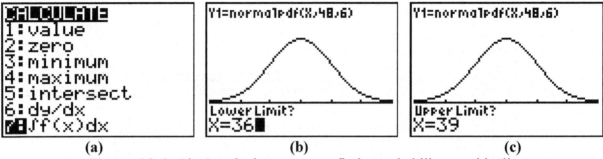

Figure 6.2.6: Alex's calculator steps to find a probability graphically

Once Alex hit enter, his calculator did something dynamic and then showed the screen depicted in Figure 6.2.7. Notice that both of Alex's methods determined that the probability of the battery failing between 36 and 39 months is about 4.4%. Given this graphical representation, Alex also interpreted that value as the area of the shaded region in Figure 6.2.7.

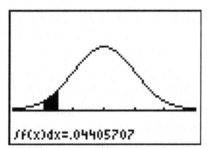

Figure 6.2.7: The probability calculated graphically

Next, Alex decided to calculate the probability that the battery fails beyond the warranty period. Since he wouldn't get any money back if this happens, he was very interested in the likelihood that the battery will fail after 42 months. He wanted to use the normalcdf command, but he needed to know both the lower and *upper* limits of that range. The lower limit is clearly 42 months.

4. What is the upper limit?

Alex realized that, in theory, the upper limit for the range of months when he wouldn't get any money back is infinite. Since there is no infinity button on the calculator, he wondered how to tell the calculator that the upper limit is without bound? The answer Alex tried is to use the largest number he can possibly tell the calculator. In scientific notation, the largest number the calculator can interpret is 1×10^{99}, which is entered in the calculator as 1E99. Alex accessed the EE command by pressing 2nd, comma. The command he used and the calculator's result are shown in Figure 6.2.8. Notice that the calculator displays "E" when the EE command is given.

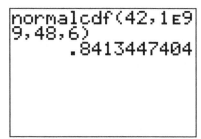

Figure 6.2.8: Alex's use of scientific notation to represent an approximation of infinity

5. What is the probability that the car battery will fail beyond the warranty period?

Alternatively, you do not need to use an absurdly large number. We know that with any normal distribution, there is less than a one in 10 thousand chance of observing a value that is more than four standard deviations above the mean.

6. What value do you obtain when you multiply the standard deviation by four and add it to the mean? Input this value as the upper limit and compare your answer to that of the previous question.

Option 7 on the CALC menu gives us a nice way to visualize the area under the curve. Let's use that feature to find the same probability. Before repeating this process, we want to clear the shading that we created. While looking at the graph, access the DRAW menu by pressing 2nd PRGM, and then select option 1, ClrDraw. This will clear the shaded region, but also redraw the normal curve. Now that your graph is displayed, we can try to calculate the probability: access the CALC menu by pressing 2nd TRACE, option 7, and then enter your lower and upper limits.

7. Did you get the same answer? Why or why not?

Option 7 under the CALC menu will only allow you to enter a lower and upper limit within the viewing window of your graph. Try to use this technique to find the probability that the car battery will fail between 42 months and the maximum *x*-value in your window.

8. How does that compare with the value you found between 42 months and infinity when you used the normalcdf command?

9. Can you explain why the probabilities are so close?

Use either technique (normalcdf or option 7) to find the probabilities for all the ranges in the warranty period. Record your answers in a table similar to Table 6.2.2.

10. Do your values make sense?

11. What happens when you add all the probabilities together? Why?

Probability of battery failure at time T
$P(T < 36) \approx$
$P(36 < T < 39) \approx 0.044$
$P(39 < T < 42) \approx$
$P(T > 42) \approx$

Table 6.2.2: Recording the probabilities of battery failure at time T

12. Since Alex intended to keep his car for five years, what is the probability that the battery will fail before 60 months has passed?

13. Do you think he will need to buy another car battery before then? Explain your reasoning.

6.2.2 Expected Value of the Warranty

Once we know the probabilities for each range, we can calculate the expected value of the warranty. Recall that the *expected value* of a random variable is the long-term average value. So, the expected value of a battery warranty is the long-term average amount of money that a customer gets back when the battery fails. It is calculated by multiplying each probability by the amount of money you would get back and then adding them all together.

14. What is the expected value of the car battery warranty?

15. From Alex's perspective, what does that value mean?

A consumer will only purchase one car battery, in which case the notion of expected value referring to the long run average doesn't really make sense. In the long run, Alex still only has the one car battery that either fails during the warranty period or fails outside the warranty period. On the other hand, the company that sells the battery would be interested in the expected value simply because of the large number of car batteries they have sold. From the company's perspective, they would want to have the expected value of the warranty as small as possible. The warranty requires them to pay in the event of their battery's failure.

16. Suggest changes to the warranty policy to reduce the expected cost of the warranty to less than $10?

17. What happens to the average warranty costs if the company can improve the life of the battery so it lasts, on average 49 months? 50 months?

Since the average life of the car battery is 48 months and Alex intended to keep this car for five years, he was concerned about what may happen outside of the warranty period.

18. What is the probability that the battery will last to the average, that is, find $P(T > 48)$?

19. What is the probability that it will last 54 months, or $P(T > 54)$?

20. What is the probability that it will last five years (i.e., 60 months)?

Alex's sister Josephine just returned home for Thanksgiving break during her senior year in college. Her car has the same battery that Alex researched and considered buying. Her car battery is 48 months old. Based on what Alex showed her, she was concerned that her battery would not make it through her final six months of school. Alex applied the formula for conditional probability to determine the probability that a 48 month-old battery will last at least another six months.

$$P(T > 54 \mid T > 48) = \frac{P(T > 54)}{P(> 48)}$$

21. What is this probability? Does Josephine need to be concerned?

6.2.3 A Special Offer

The company that makes the batteries is considering offering a special coupon for owners of really old batteries that are approaching their end of life. They plan to offer this coupon to owners with the 5% oldest batteries. They therefore need to know the battery lifetime that there is only a 5% chance of surpassing. Conversely, 95% of the batteries will have failed before reaching this age. When given the probability, you can use your calculator to find the corresponding *value* on the normal curve. To do this, we will once again call up the distribution menu from the calculator. However, this time we will be inverting the normal distribution. The command is option 3, invNorm(, and the parameters are the percentage, the mean, and the standard deviation, respectively. Figure 6.29 shows the calculator screens.

 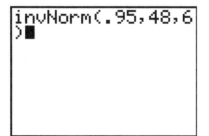 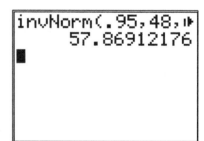

Figure 6.1.9: Using the invNorm command

There a 95% chance that the battery will fail before 57.9 months. However, the company is considering limiting the coupon to just the 1% oldest batteries.

22. There a 99% chance that the battery will fail before how many months?

23. What if government regulations forced the company to offer the coupon to the oldest 10% of the batteries. There is a 90% chance the battery will fail before how many months?

Excel equivalent to invNorm

Excel has a function that the returns the same value as this calculator function. This function, NORM.INV, is presented below.

=NORM.INV(probability, mean, standard_dev)

=NORM.INV(0.95, 48, 6)

Section 6.3 The Standard Normal Distribution

One of the important features of any normal distribution is that it can be rescaled to the *standard normal distribution*. In the standard normal distribution, the mean is zero and the standard deviation is one. The standard normal distribution allows you to make quick probability estimates, because the probabilities of being ±1, ±2, ±3 or ±4 standard deviations from the mean are well known. Those probabilities are given in Table 6.3.1. Thus, 0.68 is a good estimate of the probability of being within one standard deviation from the mean, 0.95 is a good estimate of the probability of being within two standard deviations from the mean, and 0.99 is a good estimate of the probability being within three standard deviations. The table also shows that there is very little chance of being four standard deviations or more from the mean. We can also use the symmetry of a normal distribution together with these probabilities to determine that the probability of x being between the mean and one standard deviation above the mean. This range has a probability of about 0.34 which half of the two-sided value.

Domain of x		Probability
Verbal	**Symbolic**	
Within one standard deviation of the mean	$\mu - \sigma \leq x \leq \mu + \sigma$	68.26%
Within two standard deviations of the mean	$\mu - 2\sigma \leq x \leq \mu + 2\sigma$	95.45%
Within three standard deviations of the mean	$\mu - 3\sigma \leq x \leq \mu + 3\sigma$	99.73%
Within four standard deviations of the mean	$\mu - 4\sigma \leq x \leq \mu + 4\sigma$	99.99%

Table 6.3.1: Probabilities based on standard deviations in a normal distribution

1. What is the approximate probability of x being between the mean and two standard deviations above the mean?

2. What is the approximate probability of x being between one and two standard deviations above the mean?

3. What would the probabilities for questions one and two be if the same questions were asked about values below the mean?

6.3.1 Converting Values to the Standard Normal

We will use the car battery example from the previous section to demonstrate how to use the standard normal distribution. Recall that Alex assumed that battery life is normally distributed with $\mu = 48$ months and $\sigma = 6$ months. Suppose he wanted to find the probability that the car battery would fail in the first 36 months. The first step in using the standard normal distribution is to find what is known as the *z-score* for 36 months. The z-score of any piece of data, x, is simply the number of standard deviations it is away from the mean. A z-score is found using the following formula:

$$z = \frac{x - \mu}{\sigma}$$

Thus, the z-score for the car battery failing in the first 36 months, is $z = \frac{36-48}{6} = \frac{-12}{6} = -2$.
This means that 36 months is two standard deviations below the mean.

4. How does the z-score tell you the value is *below* the mean?

We saw in the previous section that the probability was also the area under the normal curve.

5. What is the largest value a probability can take on?

6. What do you think the total area under the curve is? Explain your thinking.

We are looking for the probability that the car battery will fail in the first 36 months. This is the same as asking for the probability that a value is less than two standard deviations below the mean.

Once you have rescaled the datum to a z-score, you can now use the standard normal distribution to estimate the probability that the battery will fail in the first 36 months. Our standard normal estimate that x lies within two standard deviations of the mean is 0.95. Our estimate of the probability that z lies further than two standard deviations from the mean is 0.05. But there are two ways that the value of z could be further than two standard deviations from the mean. The value of z could be greater than the mean (z is positive), or it could be less than the mean (z is negative). Using the symmetry of the normal distribution, the 0.05 probability of being further than two standard deviations from the mean needs to be divided in half. Thus, the probability of being more than two standard deviations *above* the mean is 0.025; while the probability of being more than two standard deviations *below* the mean is also 0.025. Applying this reasoning to Alex's question, there is only about a 2.5% chance (one in 40) that the car battery would fail within the first 36 months.

7. How could you estimate the probability that the battery fails after 36 months?

The probability we found contains one of the tails of the distribution. There is slightly more work to be done if we want to estimate a probability which does not contain a tail of the distribution. For example, suppose Alex wanted to find the probability that the battery will fail between 48 and 60 months. First, he would need to convert these data points to z-scores.

8. Find the z-scores for 48 and 60 months.

Recall that the standard normal z-score is the distance from the mean of the distribution. Using the symmetry of the normal distribution, $P(z<0) = 0.5$ and $P(z>0) = 0.5$. (What is the probability that $z = 0.5$?) Now, the probability estimate that z is between -2 and 2 is 0.95. But using symmetry again, one-half of that probability is for being negative, and the other half positive. So, the probability estimate that z is between zero and two is half of 0.95, which is 0.475. So, there is almost one in two chance that the battery will fail between 48 and 60 months.

Lead Authors: Kenneth Chelst and Thomas Edwards

9. Use a calculator to sketch the standard normal distribution and shade the area that corresponds to falling between the z-scores representing battery failure between 48 and 60 months.

10. Estimate the probability that $z < 1$?

11. Estimate the probability that $z < -2$?

12. Estimate the probability that the battery will fail between 36 and 54 months?

A common error in this situation involves subtracting in the incorrect order. But there is an easy way to tell if the order is wrong. If you use the previous values and subtract them in the wrong order, your result is negative.

13. How can you tell that a negative number cannot be the correct answer to a question regarding probability (or area)?

Section 6.4 *Rappin' Skoop Dogg*: Seasonal Demand

The Pineapple Technology Corp. has developed a new item just in time for the holiday season. *Rappin' Skoop Dogg* is an upright stuffed hound dog, dressed in a leather-like jacket and pants costume. It sings and dances while standing on a replica of a stage and holding a microphone. There are ten different rap songs that *Skoop Dogg* can sing as well as dance to. Additional songs can be downloaded from the Pineapple Technology Corp. web site. There are three settings for activating the dog. *Skoop Dogg* has a motion detector and can be set to activate when anyone approaches. It can also be activated by clapping your hands or pushing a button. Commercials for *Rappin' Skoop Dogg* can be seen on the VideoTube website.

The Chillioz Toys store chain is planning on placing a large order in anticipation of the holiday season. The Pineapple Technology Corp. charges $45 per item for purchases of more than 10,000 units. Chillioz Toys plans to price *Skoop Dogg* at $99. However, items left over after the holiday season will be steeply discounted to $30. There is an inventory carrying cost of $2 per item left over after the holiday season ends. This data is summarized below:

P = Purchase price	$45
S = Selling price	$99
D = After Christmas sale price	$30
H = Inventory carry over cost	$ 2
Profit on regular sale	$99 – 45 = $54
Net cost at end of season for leftovers	$30 – 45 – 2 = -$17

Based on past experience with similar products, executives at Chillioz Toys know there is high potential, but also a great deal of uncertainty. They forecast that the demand for this item is normally distributed with mean of 21,000 and standard deviation of 4,200 (one-fifth of the mean).

Roy Charles, VP for marketing, is interested in analyzing the impact of different ordering policies. He starts by wondering how many, X, to order so that there is only a 40% chance that the demand will be less than this number. He therefore wants to determine the value of *x* such that:

$P(X < x) = 0.40$

To determine this value, VP Charles uses the ***inverse of the normal distribution***. invNorm(is one of the functions that can be found in the DISTR menu of a graphing calculator. The syntax is invNorm(x, μ, σ). Figure 6.4.1 contains a calculator screen for this calculation. From this calculation, VP Charles can see that there is about a 40% chance that the demand for *Rappin' Skoop Dogg* will be less than 19,936 units.

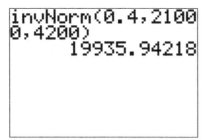

Figure 6.4.1: Using the inverse of the normal distribution

VP Charles noticed that the profit margin was much larger than the potential loss if a product is not sold before the end of the Holiday season. He is considering ordering more than the average demand. This time, he wants to find out how many to order so that there is only a 40% chance that the demand will be *more* than this number.

$P(X > x) = 0.40$

However, the calculator only provides the left hand tail of the normal probability density function. He must first apply the principle of the ***complement***.

$P(X > x) = 1 - P(X < x)$

If the area to right of x is going to be 0.4, the area to the left will have to equal 0.6.

1. Use the inverse normal function to determine how many toys to order.

VP Charles wants to look at the decision from a different point of view. He is hoping to sell as many units as possible and does not want to run out of stock. He considers placing an order that is 1,000 units more than the average. He wishes to know the probability that the demand will be more than that amount.

Recall that in Section 6.2, we graphed a normal distribution and used the definite integral feature of the calculator (option 7 in the CALC menu) to find the probability that a battery would fail between 36 and 39 months after purchase (see Figures 6.2.6 and 6.2.7 for details). Using the same method, we now find the probability that demand for the *Skoop Dogg* toy will be more than 1,000 units greater than the mean. In other words, VP Charles wants to know the probability that demand will exceed 22,000 units. To answer this question, the calculator screens shown in Figure 6.4.2 display the calculator set-up. For the definite integral the lower limit should be set to 22,000 and the upper limit set to Xmax.

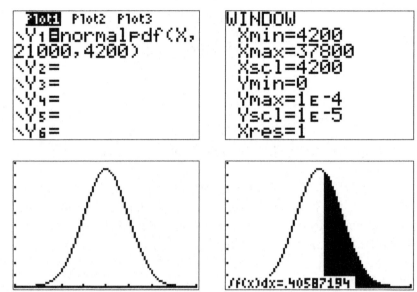

Figure 6.4.2: Screenshots for finding probability using the graph normal pdf graph

1. What is the probability that demand exceeds 22,000 units?

Until now VP Charles focused just on the uncertainty of the demand and not on profits. He considers two scenarios.

Scenario I: He orders only 21,000 units but the demand was actually 22,000

In this case Cheelioz would sell out. The total profit is just (21,000)($54)= $1,134,000 However, the company lost 1,000 potential sales and missed out on $54,000 extra profits.

Scenario II: He orders 22,000 units but the demand was actually 21,000.

In this case Cheelioz would not sell out. They would have to discount 1,000 units after the season. The total profit is just (21,000)($54) – (1,000)($17) = $1,117,000.

6.4.1 Two Policies: Simulated Demand

VP Charles is not sure how to analyze the decision. How does he balance the risk of having too few against the risk of having too many? He decides to ask two assistant vice presidents, Tom Kawtion and Daniella Egr, to make recommendations. Tom suggests a conservative policy of ordering 22,000 *Rappin' Skoop Dogg* toys. Daniella is more optimistic and suggests an order of 26,000. Roy Charles attempts to evaluate these two policies by simulating ten random seasons of demand. Table 6.4.1 presents an analysis of Tom's policy with the randomly generated demand listed in the second column. The third column will record how many items would have been sold at full price. For example, the first simulated demand was only 14,092. This was less than the amount ordered. As a result only 14,092 would be sold at full price. The rest would be marked down at the end of season, and that number will be reported in column four. In the last column the number sold at full price will be multiplied by $54. This profit is reduced by all of the marked down items. The number of marked down items is multiplied by the net loss of $17 and the total is subtracted from the profit. The second simulated demand was 23,447. This is greater than the amount ordered. Thus the number that could be sold at full price is just 22,000, the number ordered. The number of marked down items would then be zero.

Tom's policy – Order 22,000 units				
Trial	Simulated demand (units)	*Rappin' Skoop Doggs* sold		Profit ($)
		Full price (units)	Discounted (units)	
1	14,092	14,092	22,000 – 14,092 = 7,908	(14,092)(54) – (7,908)(17) = 626,532
2	23,447	22,000	0	(22,000)(54) = 1,188,000
3	20,152	20,152		
4	18,917	18,917		
5	28,899	22,000		
6	13,100	13,100		
7	25,176	22,000		
8	22,283	22,000		
9	17,613	17,613		
10	23,450	22,000		
Average	20,713	19,387		

Table 6.4.1: Simulated demand and a conservative order policy

The details in Table 6.4.1 could be worked out by hand, but we will use the LIST feature of the calculator to facilitate the task. To access LISTs, from the STAT menu select EDIT. This brings up L1, L2, and L3 as shown in Figure 6.4.3. Scrolling to the right will reveal L4, L5, and L6. Now, follow the steps below to create a finished table in L1 through L4.

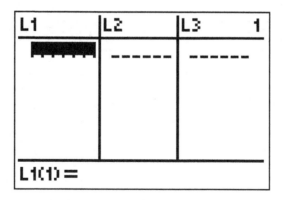

Figure 6.4.3: Accessing L1 through L3

- Enter the simulated demands from Table 6.4.1 into L1 (see Figure 6.4.4(a)).

- Enter into L2 the simulated demand if it is less than or equal to 22,000, and 22,000 if the simulated demand is greater than 22,000. L2 now contains the number that would be sold at full price for each simulated demand (see Figure 6.4.4(b)).

- In L3, we want to place the number of units that would be discounted. For each line, that number is 22,000 minus the number sold at full price. We can program the calculator to do these calculations by scrolling to L3, pressing the up-arrow until L3 is highlighted, then typing: 22000 – L2 and pressing ENTER (see Figure 6.4.4(c) & (d)).

- L4 will contain the calculations necessary to find the profit for each simulated demand. Again, we can program the calculator to do so. Scroll to L4, press the up-arrow until L4 is highlighted, type 54 * L2 – 17 * L3, and press enter (see Figure 6.4.4(e) & (f)).

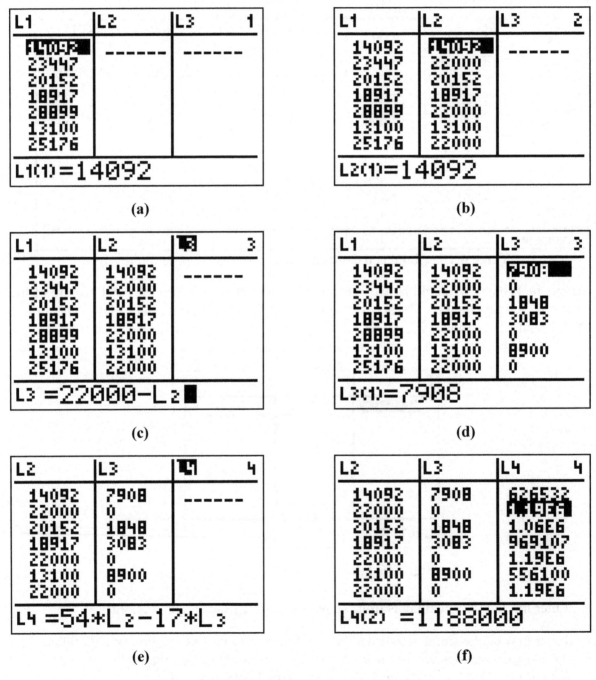

Figure 6.4.4: Using LISTSs to complete Table 6.4.1

When you have completed Table 6.4.1, either by hand or using the LIST feature of the calculator, answer the questions that follow.

2. How many times did the demand exceed the order policy? In each case what value is recorded in the column marked discounted and why?

3. What was the average demand for the simulated data? How close is it to the forecasted average? Are you surprised that the number is several hundred units below the forecasted average?

4. What were the least and most profits earned?

5. Why does the exact same total profit of $1,188,000 appear several times?

Table 6.4.2 is set to repeat the above analysis for Daniella's order policy. You are asked to help Roy Charles complete the calculations for trials 2 thru 10, either by hand or with an assist from the calculator.

	Daniella's policy – Order 26,000 units			
Trial	Simulated demand (units)	*Rappin' Skoop Doggs* sold		Profit ($)
		Full price (units)	Discounted (units)	
1	14,092	14,092	26,000 – 14,092 = 11,908	(14,092)(54) – (11,908)(17) = 558,532
2	23,447			
3	20,152			
4	18,917			
5	28,899			
6	13,100			
7	25,176			
8	22,283			
9	17,613			
10	23,450			
Average	20,713			

Table 6.4.2: Simulated demand and an optimistic order policy

6. How many times did the demand exceed the order policy?

7. What were the least and most profits earned?

8. How do these profits compare to the results for Tom's order policy.

9. Why is the maximum total profit not repeated several times?

10. What is the average profit under Daniella's Policy?

11. Which policy would you recommend to VP Roy Charles?

12. Which student policy earned the most profit?

13. What was the average profit?

6.4.2 Excel alternative approach

In this section we demonstrated the list function within a graphing calculator as we analyzed a series of randomly generated demand points. Excel can naturally and more efficiently handle all of the steps presented in Figure 6.4.4

Tom's Policy – Order 22,000 units				
Trial	Simulated demand (units)	*Rappin' Skoop Doggs* sold		Profit ($)
		Full price (units)	Discounted (units)	
1	14,092	14,092	7,908	626,532
2	23,447	22,000	0	1,188,000
3	20,152			
4	18,917			
5	28,899			
6	13,100			
7	25,176			
8	22,283			
9	17,613			
10	23,450			
Average	20,713			

Table 6.4.3: Excel simulated demand and a pessimistic order policy

In this spreadsheet, the data and calculations for trial 1 are in row 5. The simulated demand of 14,092 is stored in cell B5. The number of items sold at full price and recorded in cell C5 is determined by using an IF statement.

=IF(B5 < 22000, B5, 22000)

This statement states compares the demand stored in cell B5 to 22,000. If the value is less than 22,000 then use that number as the total sold at full price. However, if this is not true and the demand exceeds 22,000, use the number 22,000 as the amount sold at full price. The copy and paste command allows you to replicate this formula down column C. The number of sales that are discounted in trial 1 (row 5) is calculated with the following formula and stored in cell D5.

=22000 – C5

Lastly, the total profit and reported in column E is determined with the following formula.

=C5 * 54 – D5 * 17

The copy and paste commands can then fill in all of the other calculations. This process can be easily replicated with any policy and any set of randomly generated demand values.

6.4.3 Class Simulation Demand

VP Roy Charles volunteered to give a lecture on demand and inventory management in a business school course. He decides to use this opportunity to explore other possible order strategies. He also wants to use a larger random sample. In teams of two, he asks every pair of students to come up with their recommended order policy in units of 100. They record that value in a copy of the following table. Next, he is going to simulate on his calculator a random number based on the normal distribution with a mean of 21000 and a standard deviation of 4200. After he calls out the demand and records it on the board, each team is to calculate its profit for that simulation. At the end of the twenty simulations, the team with the highest average profit will receive a free *Rappin' Skoop Dogg*.

Trial	Student's Policy – Order _____ units			
	Simulated demand (units)	*Rappin' Skoop Doggs* sold		Profit ($)
		Full price (units)	Discounted (units)	
1				
2				
3				
4				
5				
6				
7				
8				
9				
10				
11				
12				
13				
14				
15				
16				
17				
18				
19				
20				
Average				

Table 6.4.4: Simulated demand and student order policy

14. Which student policy earned the most profit?

15. What was the average profit?

6.4.4 A Formula for Ordering

After class, the instructor Dr. Regina Smarty whispered to the VP that operations researchers have developed a mathematical formula to determine the optimal order policy. This policy is optimal on average. The formula uses the selling price (S), purchase price (P), discounted price (D), and inventory carrying cost (H).

According the formula the optimal value of x is such that

$$P(X < x) = \frac{S-P}{S-D+H}$$
$$= \frac{99-45}{99-30+2}$$
$$= \frac{54}{71}$$
$$\approx 0.761$$

In this case, the optimal policy is the value of x such that the probability that demand is less than x is 0.761. If no integer value of X exactly matches this probability, the formula states to pick the next largest integer.

   16. Using the invNorm(function, what is the optimal number to order?

   17. How does this value compare to the best policy found by the students?

VP Charles did not have time to check out the claims of Dr. Smarty. On his way back to the office, he was thinking of calculating the average profit for this formula derived policy. He was going to use the class-simulated data in Table 6.4.3.

   18. Why might the best student's order quantity have produced a higher average profit than the order quantity based on the formula?

Chapter 6 (Normal Distribution) Homework

1. An expert witness in a paternity suit testifies that the length of human gestation is approximately normally distributed with parameters $\mu = 270$ days and $\sigma = 10$ days. The defendant in the suit is able to prove that he was out of the country during a period that began 290 days before the birth of the child and ended 240 days before the birth. If the defendant was, in fact, the father of the child, what is the probability that the mother could have had the very long or very short gestation indicated by the testimony? (This problem is adapted from "A first course in probability" book by Ross.)

2. Assume that the life of a laptop computer has a normal distribution with a mean of 38 months and a standard deviation of eight months.
    a. Find the probability that a randomly selected laptop will need to be replaced in less than 42 months.
    b. What should the warranty period be to replace a malfunctioning laptop if the company does not want to replace more than 5% of all laptops sold?

3. Factory tests of the life of tires indicated that tire life is normally distributed with a mean mileage of 50,000 and a standard deviation 3,000 miles.

    a. What is the probability that the life of a random tire will exceed 55,000 miles?
    b. What is the probability that the life of a random tire will not reach 47,000 miles?
    c. The factory wants to set the guaranteed mileage so that no more than 10% of the tires will have to be replaced. What guaranteed mileage should the factory announce?

4. Graduate Management Aptitude Test (GMAT) scores are widely used by graduate schools of business as an entrance requirement. Suppose that in one particular year, the mean score for the GMAT was 476, with a standard deviation of 107. Assuming that the GMAT scores are normally distributed, answer the following questions:
    a. What is the probability that a randomly selected score from this GMAT falls between 476 and 650.
    b. What is the probability of receiving a score greater than 750 on a GMAT test?
    c. What is the probability of receiving a score of 540 or less on a GMAT test?
    d. What is the probability of receiving a score between 330 and 440?
    e. If a business school wants to consider the top 5 percent students based on their GMAT scores, which score should be considered as a minimum score?

5. After each surgery, a surgeon has appointments with her patients to make sure that the healing process is proceeding well. The duration of each appointment is assumed to be normally distributed with a mean of 10 minutes and a standard deviation of three minutes.
    a. Find the probability that a randomly selected appointment takes less than 15 minutes.

b. Assuming independent appointment times, what is the probability that at least one of the doctor's 20 appointments will be longer than 15 minutes? (Hint: Use the concept of a complement.)

c. If the surgeon works 210 minutes before his lunch, what is the probability of completing 20 patient visits before her lunch?

6. The target value for the thickness of a machined cylinder is 8 cm. The upper specification limit is 8.2 cm and the lower specification limit is 7.9 cm. A machine produces cylinders that have a mean thickness of 8.1 cm and a standard deviation of 0.1 cm. (adapted from http://stat.wvu.edu/srs/modules/Normal/diameters.html)
    a. What is the probability of producing a cylinder within the limits?
    b. What is the probability of exceeding the upper limit?

If the thickness of a cylinder is too thin and less than the lower limit, the product cannot be used. It is scrapped and the company has wasted $6. However, when the cylinder produced is too thick, the cylinder can be used with some rework. The cost of reworking the cylinder is $1.50.
    c. What is the expected cost associated with producing cylinders that are not within specifications?

7. Suppose you must establish regulations concerning the maximum number of people who can occupy an elevator. You decide to assume that the weight of one person chosen at random follows a normal distribution with a mean of 70 kg and a standard deviation of 20 kg. What's the probability that the total weight of eight people exceeds 600 kg?

8. An assembly line consists of six stations in order to produce a special type of machine. The task at each station takes 70 minutes on average and has a 20 minute standard deviation.
    a. What is the probability that a machine can be assembled in eight hours of work?
    b. Assume we add a worker to each station in order to reduce the average time of each station to an hour. The standard deviation is assumed not to change. Now what is the probability of finishing a machine in 8 hours of work?

9. Workers A and B are responsible for fixing products returned by customers. Suppose each starts his day with 13 products to repair. Assume the time they spend to fix each product follows normal distribution with the parameters given in Table 1.

Worker	Mean (minutes)	Standard Deviation (minutes)
A	30	10
B	40	5

Table 1: Parameters of normal distribution for time to fix a product

   a. Which employee has more variability in his work?
   b. Which one is preferred by a manager who favors consistency?
   c. What is the probability of Worker A finishing all the products in 7 hours?
   d. What is the probability of Worker B finishing all the products in 7 hours?

11. Ms. Kuo has had significant stomach pain for the last two days. She scheduled an appointment at the Dreyer Urgent Care Center for today. When she goes to the center, she has to first register at the reception desk. When the appointment time comes, a nurse will call her and do pre-screening before the doctor starts examining her.

The duration of each of the tasks is normally distributed. The mean duration and standard deviation for each task is given in Table 2. Since Ms. Kuo is in significant pain, she is eligible for the fast track. She does not have to wait between these tasks.

Task	Mean (minutes)	Standard deviation (minutes)
1) Patient registration	3.45	0.85
2) Nurse primary screening (triage)	6.40	2.00
3) Medical Doctor thorough examination	15.90	5.60

Table 2: Parameters of normal distribution for time to process a patient

a. What is the probability that the registration will take more than 3 minutes?
b. Her appointment is at 9:00 a.m. Since the registration takes some time, she has to arrive at the registration desk earlier than the appointment time. To be 90% confident that she will be on time for her appointment, how early should she arrive at the registration desk?
c. What is the probability that the triage will take less than 4 minutes?
d. What is the probability that the registration and the triage will take less than 8 minutes?
e. Calculate the average stay in the Dreyer Urgent Care Center, including the registration, triage, and examination.
f. What is the probability that she will finished with her examination in less than 30 minutes?

12. After the examination, the doctor may ask her to have a lab test or an X-ray scan, or both. If he does that, he gets the test results soon afterwards. The doctor meets with the patient to review the test results.

Task	Mean (minutes)	Standard deviation (minutes)
1) Lab test	13.67	5.3
2) X-ray room	20.70	6.5
3) Doctor meeting to discuss test results	7.58	2.5

Table 3: Parameters of normal distribution for the time of test related tasks

a. The doctor has sent Ms. Kuo to the lab. What is the probability that she will be back within 10 minutes?
b. The doctor thinks he can thoroughly examine one patient until Ms. Kuo comes back from the lab. The doctor will see only one patient during this time even if Ms. Kuo is not yet back from the lab. Assume Ms. Kuo's lab tests took 16

minutes to complete and she is returning to the doctor's office. What is the probability that Ms. Kuo will wait at most one minute for the doctor?

c. The doctor has started the thorough examination of that next patient. What is the probability that after finishing this next patient, he will wait for Ms. Kuo more than 2 minutes to return from the lab?

d. Ms. Kuo has a friend, Hatice Myer, who works fifteen minutes from the clinic. She has agreed to leave work and pick her up after the examination. Ms. Kuo is trying to decide what time she should tell her to come. Describe the basic tradeoff that Ms. Kuo must take into account in specifying the time? What are the uncertainties to take into account?

Ms. Kuo talked to the receptionist as to how frequently patients require additional tests. The receptionist's estimates are provided in Table 4.

Alternative	Possible Tests	Chances
1	No test	40%
2	Lab test	30%
3	X-ray	15%
4	Lab test and X-ray	15%

Table 4: Percentage of patients requiring different types of tests

e. For each of the four alternatives, calculate the mean total time spent in the Urgent Care and its standard deviation. Record your answers in Table 5. Do not forget that the doctor will meet with Ms. Kuo a second time if tests are required.

	Total Time Spent at Urgent Care	
Possible Tests	Mean	Standard Deviation
No test		
Lab test only		
X-ray only		
Lab test and X-ray		

Table 5: Time spent in urgent care for patients requiring different types of tests

f. Ms. Kuo is considering asking her friend to come by at 9:45 a.m to pick her up after completing her 9:00 a.m. appointment. Calculate the probability of finishing by 9:45 a.m. for each of the four alternatives: No test, Lab test, X-ray, and Lab test and X-ray.

Possible Tests	Probability of being ready to leave by 9:45 a.m. for different alternatives
No test	
Lab test	
X-ray	
Lab test and X-ray	

Table 6: Calculate probability of being ready to leave by 9:45 a.m.

g. If Ms. Kuo does not require any tests, she will likely to wait to be picked up at 9:45 a.m. What is the probability that she will have to wait at least 10 minutes if she has no tests?
h. Now consider all of the possible situations. If her friend Hatice comes exactly at 9:45 a.m., how likely will Hatice end up waiting more than 10 minutes? (Hint: You will need to determine the weighted sum of the individual alternatives. The weights are the likelihood of each alternative.)

Possible Tests	Chances	Probability of departure later than 9:55 a.m. in each event
No test	40%	
Lab test	30%	
X-ray	15%	
Lab test and X-ray	15%	
Total Probability - weighted sum		

Table 7: Calculate probability of driver waiting more than 10 minutes for patient

13. Total cholesterol is assumed to be normally distributed with a mean 5.21 mmol/L and a standard deviation of 1.03 mmol/L. The medical interns are observing the total cholesterol levels of patients in a research study. The Hyperlipidaemia Society classifies the total cholesterol level as,
    - desirable (<5.2 mmol/L)
    - borderline (5.2 to 6.4 mmol/L)
    - abnormal (6.4 to 7.8 mmol/L)
    - high (>7.8mmol/L)

    a. What percent of the patients have cholesterol that is in the desirable range?
    b. What percent of the patients have cholesterol levels that are borderline?
    c. What percent of the patients have abnormal cholesterol levels?
    d. What percent of the patients have high cholesterol?

14. A pharmacist is selling two types of cholesterol lowering drugs, Drug *A* and Drug *B*, and he is considering the stocking levels of each drug. From his records, the demand for Drug *A* is 20 boxes with a standard deviation of five boxes and the demand for Drug *B* is 10 boxes with a standard deviation of three boxes a month. Both are independent and normally distributed.
    a. How many boxes of each drug should he have at the beginning of each month if he wants to fill 90% of each demand?
    b. The combined monthly number of patients with high cholesterol in the area hospital is normally distributed with a mean of 30 and a standard deviation of 5.8. The hospital is considering using Drug *C* for lowering cholesterol. It is a substitute for both Drug *A* and Drug *B*. How many boxes of Drug *C* should be kept at the beginning of each month if the hospital wants to fulfill 90% of the patient requirements?

c. Compare the number of Drug C boxes to the total number of boxes of Drug A and B. Explain why the hospital needs more or less boxes of Drug C to satisfy 90% of the demand?
d. The prices for Drug A, B, and C are $44, $40, and $47, respectively. If the hospital wants to fulfill 90% of the patient requirements, which option is more cost effective, stockpiling both Drug A and Drug B, or stockpiling Drug C?
e. At what price for Drug C would a manager be indifferent between stocking just Drug C or keeping both Drug A and B?

15. Recent reports have shown that there is an inverse relationship between calcium intake and body weight. Thus, changing the calcium intake may reduce the prevalence of obesity. An experiment, done on middle-aged women. It was found that if the ratio of calcium intake to protein intake is 9 mg to 1 g, then the average yearly weight change will be $\mu_1 = 0.43$ kg with a standard deviation of $\sigma_1 = 0.61$ kg. The weight gain is assumed to be normally distributed.
    a. What is the probability that a middle-aged woman will lose weight in a year if her calcium to protein intake ratio is 9 mg:g?
    b. What is the probability that she will gain more than one kilogram per year if her calcium to protein intake ratio is 9 mg:g?

If the ratio of calcium to protein intake is 20 (mg:g), then the average weight change will be $\mu_2 = -0.01$ kg/year with a standard deviation of $\sigma_2 = 0.57$ kg/year.
    c. According to this given information, how likely is a middle-aged woman to lose weight in a year if her calcium to protein intake ratio is 20?
    d. What is the probability that she will gain more than 1 kg/year if her calcium to protein intake is 20?
    e. Which calcium to protein ratio is more helpful to a person who wants to lose weight and keep fit?

16. There are various intervention methods to reduce the impact of an influenza pandemic. Among those, vaccination is the most effective one. It reduces both morbidity and mortality. However, a vaccine might not be available in the initial stages of the epidemic. It takes on average 45 days to develop a vaccine. It takes another 15 days until it is readily available. Therefore, there is value in slowing the spread of the disease until the vaccine is available.

The peak of the pandemic can be delayed by other effective intervention methods such as use of antiviral drugs, self isolation, and contact reduction. Very sick people should be isolated from other people. Contact reduction is ensured by closure of schools and day care centers. Antiviral treatment is given to 10% of the patients, who are very sick.

The time until the peak spread of the pandemic after its emergence is assumed to be normally distributed. Table 8 provides estimates of the mean number of days and standard deviation until the pandemic reaches its peak.

	μ (days)	σ (days)
No Intervention	77	26
Isolation	45	15
Contact reduction	52	17
Antiviral Treatment	53	18
All interventions combined	39	13

Table 8: Peak of the epidemic with different intervention methods

a. In how many days, does the pandemic reaches its peak on average if no intervention is used?
b. In how many days, will the vaccine be readily available?
c. What is the probability that the pandemic reaches its peak before the vaccine is readily available if no intervention is used?
d. What is the probability that the pandemic reaches its peak before the vaccine is readily available if antiviral treatment is used?
e. Even though the antiviral treatment is the second most effective tool for delaying the peak, why do you think other intervention methods might be preferred?
f. How many days does self-isolation delay the peak of the epidemic on average compared to no intervention?
g. Do you think contact reduction considerably slows down the spread of the epidemic?
h. Why is the impact of all interventions combined not equal to the sum of other interventions?
i. What is the probability that the pandemic reaches its peak before the vaccine is readily available if all interventions are used simultaneously?

Chapter 6 (Normal Distribution) Summary

What have we learned?

The normal distribution is a mathematical model for random variables whose values tend to cluster around the mean. The distribution's bell-shaped curve is well known and has many useful mathematical properties. The area under the curve between any two values corresponds to the probability that the actual value of the random variable is between those two values. Technological tools play a large role in putting the normal distribution to work. We used the probability and statistics features of a graphing calculator as our technological tool.

Once we understood the differences between the Normalpdf, Normalcdf, and invNorm functions on a graphing calculator, we were able to use the:
- Area under the graph of the Normalpdf to find probabilities.
- Normalcdf command to obtain probabilities directly.
- invNorm command to find the corresponding value of x for a given probability.

When we were able to find normal distribution probabilities, we used:
- Theoretical probabilities to determine expected values of various policies, and then make policy decisions.
- Probabilities to predict future behavior.
- Probabilistic simulations to investigate the economic impact of policies.

Terms

Cumulative distribution function   A cumulative distribution function (cdf) is a function that describes the likelihood of a random variable X having a value less than x. This corresponds to finding the area under the pdf everywhere to the left of x.

Inverse of the normal   The invNorm command on the graphing calculator is the inverse of the normal cdf. Where the cdf gives the probability that a score is between any two values, the inverse of the normal provides the value of X such that $P(X < x) = p$ for a given probability, p.

Mean   The mean is a measure of central tendency. The mean of a sample is denoted by $\bar{x}$, and the mean of a population is denoted by μ. The mean of a set of data can be calculated using the formula:

$$\bar{x} = \frac{\sum_{i=1}^{n} x_i}{n}.$$

Normal distribution   The normal distribution is a continuous probability distribution. This distribution is appropriate for variables that tend to cluster around the mean. The normal distribution is a good approximation for many random variables that are the sum or mean of a collection of other random variables.

Parameter   Parameters are statistics that characterize a distribution. All random variables that are distributed normally can be modeled by a bell-shaped curve. The shape and size of that curve are determined by the mean and standard deviation of the random variable. The mean and standard deviation are the parameters of the normal distribution.

Probability density function   A probability density function (pdf) is a function that describes the relative probability of a random variable taking on a specific value. The area under the graph of a pdf gives the probability that the value of the associated random variable lies between the left- and right-hand bounds of the region. The total area under a pdf curve is always equal to one. The normal pdf is a bell-shaped curve.

Range   The range is a measure of variability or spread. The range of a set of data can be calculated by subtracting the smallest value from the largest value.

Standard deviation The standard deviation is another measure of variability. The standard deviation gives an indication of how closely the data are packed together. The standard deviation of a sample is denoted by S_x, and the standard deviation of a population is denoted by σ. The standard deviation of a set of data can be calculated with the following formula.

$$S_x = \sqrt{\frac{\sum_{i=1}^{n}(\bar{x}-x_i)}{n-1}}$$

Variance The variance is another measure of variability. It describes how far values of a random variable are from the mean. Variance is a useful statistic when taking the sum of two or more normally distributed random variables. The variance is the square of the standard deviation.

Chapter 6 (Normal Distribution) Objectives

You should be able to:

- Recognize when the normal distribution is an appropriate model for a random variable.
- Calculate the probability that the value of a normally distributed random variable is less than a given value.
- Calculate the probability that the value of a normally distributed random variable is greater than a given value.
- Calculate the probability that the value of a normally distributed random variable is between two given values.
- Calculate the value, X, of a normally distributed random variable so that $P(X < x) = p$, for some given probability, p.
- Use real or simulated data to perform economic analyses.
- Calculate probabilities when two normally distributed random variables are added to together.
- Calculate probabilities when n random variables are added to together.

Chapter 6 Study Guide

1. What is a random variable?

2. What is a continuous random variable?

3. What makes the normal distribution different from other distributions you have studied?

4. Explain why it makes sense that the mean, median, and mode of a normally distributed random variable are all equal.

5. How is the normal probability density function (normalpdf) useful?

6. How is the normal cumulative distribution function (normalcdf) useful?

7. How is the inverse of the normal function (invNorm) useful?

8. What is the purpose of the standard normal distribution?

9. When two normally distributed random variables are added, the mean of the new distribution is the sum of the means of the original two. The standard deviation of the new distribution is, however, not the sum of the standard distributions of the original two. Why not?

Made in United States
North Haven, CT
15 July 2022